DANCING COCKATOOS
AND THE DEAD MAN TEST

DANCING COCKATOOS
AND *THE* DEAD MAN TEST

How Behavior Evolves
and Why It Matters

MARLENE ZUK

W. W. NORTON & COMPANY
Celebrating a Century of Independent Publishing

For information about permission to reproduce selections from this book, write to Permissions, W. W. Norton & Company, Inc., 500 Fifth Avenue, New York, NY 10110

For information about special discounts for bulk purchases, please contact W. W. Norton Special Sales at specialsales@wwnorton.com or 800-233-4830

Manufacturing by Lakeside Book Company
Production manager: Devon Zahn

Library of Congress Cataloging-in-Publication Data

Names: Zuk, M. (Marlene) author.
Title: Dancing cockatoos and the dead man test : how behavior evolves and why it matters / Marlene Zuk.
Description: First edition. | New York, NY : W. W. Norton & Company, [2022] | Includes bibliographical references and index.
Identifiers: LCCN 2022008968 | ISBN 9781324007227 (cloth) | ISBN 9781324007234 (epub)
Subjects: LCSH: Animal behavior. | Animal intelligence.
Classification: LCC QL751 .Z85 2022 | DDC 591.5—dc23/eng/20220524
LC record available at https://lccn.loc.gov/2022008968

ISBN: 978-1-324-06440-4 pbk.

W. W. Norton & Company, Inc., 500 Fifth Avenue, New York, N.Y. 10110
www.wwnorton.com

W. W. Norton & Company Ltd., 15 Carlisle Street, London W1D 3BS

1 2 3 4 5 6 7 8 9 0

Contents

Introduction

A colleague of mine used to have two stock answers when the passenger next to him on an airplane asked what he did for a living. If he was happy to talk to the person, he said he was a marine biologist, which spurred all kinds of inquiries about ocean animals and anecdotes about snorkeling vacations. If he wasn't interested in conversation, he said he was an evolutionary biologist, which usually resulted in the person smiling tightly and returning to their electronic device.

I'm usually not quite so premeditated about my answer to that question, though I have noticed that if I say I study insects (which is true), people will ask me how to kill the pests in their garden, or maybe about beekeeping, which I know nothing about. If, on the other hand, I say that I work on animal behavior (which is also true), I get lots of stories about people's pets and questions about the latest nature documentaries. They often want to talk about how much smarter animals are than we give them credit for, although admittedly that usually goes for animals other than insects. Either way, I like talking to people about animals, and I am always interested in hearing how they interpret the behavior of the creatures around them.

Often, the conversation turns to how much animals are like

humans, though again this topic usually excludes insects. People
have always wanted to draw parallels between our lives and those
of animals. Author Helen Macdonald eloquently points out, "The
deepest lesson that animals have taught me: how easily and uncon-
sciously we see other lives as mirrors of our own."[1] Of course, as she
goes on to say, "Animals don't exist in order to teach us things, but
that is what they have always done, and most of what they teach us
is what we think we know about ourselves."

At the same time, what we know about ourselves these days is continually infused with the latest discoveries about where our behavior
comes from, and particularly about how so many of our most human
characteristics are innate, and have a basis in our genes. Headlines
are full of declarations like: "Our Politics Are in Our DNA," or "To
Move Is to Thrive. It's in Our Genes." A paper in the prestigious scientific journal *Proceedings of the National Academy of Sciences* was titled
"Trust Is Heritable, Whereas Distrust Is Not."[2] Even dog ownership
was said to have a genetic basis.

Britain's *Telegraph* screamed in June 2019: "There Is No Point
Sending Your Children to Eton Because Education Is in Your Genes,
Says Geneticist." The article was reporting on work by Robert
Plomin, who also said that while "environmental influences are
important, too . . . they are largely unsystematic, unstable and idio-
syncratic." Sarah Knapton, the newspaper's science editor, flatly
stated that Plomin, a professor of behavioral genetics at King's
College London, had shown that "academic success is written in
the genes." Not "suggested," mind you, "shown." Even the #MeToo
movement is, at its heart, disputing the long-held notion that men
are "naturally" inclined to hit on women, and whether there is anything we can really do about it. And when people talk about what's
natural, or inherent, or fixed, or instinctive, to use just a few of the
synonyms that are bruited about, they mean what biology, or our
genes, determines.

By contrast, others claim the environment is all powerful. Prominent "functional medicine" practitioner, Chris Kresser, writes in

his blog, "At one time scientists believed our DNA held the key to preventing and reversing disease. But we now know that our environment—not our genes—is the primary driver of health and longevity."[3] He discusses the "exposome," the lifetime sum of an individual's environmental exposures, as critical in determining health and disease. And to many people, one can make an uncomplicated distinction; according to a website maintained by no less august a group than the American Association for the Advancement of Science (AAAS), "Some behavior, called innate, comes from your genes, but other behavior is learned, either from interacting with the world or by being taught."[4] It then presents examples of pet behavior, neatly divided into "learned" or "innate."

The Greek Not-So-Blank Slate

All of this brouhaha about what's learned and what's cultural is, of course, the old nature-nurture debate. Questions about what we learn and what we inherit have long dominated our public conversations about equality, about why our talents and predilections vary among us and where they come from. Are we blank slates, or are our lives and actions determined long before we are born, by our genes and their effects on our bodies and brains? People have been trying to answer this question for centuries if not longer, and the question has existed in some form even before we knew there was such a thing as a gene. Scholars have long argued about the nature of "innateness," a concept that, as British philosopher Richard Samuels says, "has led a busy life." Philosopher Paul Griffiths, whose work I will return to in chapter 10, suggests that we have a folk concept of innateness that we use when we think about what living things are like. Both scientists and the public want to weigh in and take sides, and they will often reach to unlikely sources for support. As columnist George Will wrote in 2019, "In the argument about which is primary, nature or nurture, the former receives an emphatic affirmation from the Founding Fathers' philosophy. Beneath the myriad patinas of culture, there is a fixed human nature that neither improves nor regresses."[5]

Even Aristotle and Plato have roles here, though the question at hand was not where our behavior comes from, but the closely related issue of how knowledge itself came to be. According to Aristotle, we use our senses to comprehend the world, which means that learning, or the environment, was sufficient to explain how we operate. Plato disagreed, claiming that we do not acquire information about the world but instead are endowed, or born, with it. Prominent philosophers have weighed in on one side or the other over the ensuing centuries, but one thing remains clear: we haven't got this question sorted out by a long shot.

The ancient Greeks had no notion of genes, of course, though they obviously knew that parents passed on characteristics to their offspring. And indeed, the argument doesn't depend on genes. Psychologists, particularly those interested in child development, have essentially renamed the conflict as the nativist-empiricist debate, which as far as I can tell is the same old genetic versus environmental wine in new bottles, though the focus isn't on DNA so much as on how human development uses information from different sources. The argument is also related to the idea of essentialism, which I will discuss in more detail in later chapters. Briefly, from the essentialist viewpoint every being has an essence that it carries with it and that transcends its experiences and other changes, such as growth and development. Worms are thus always wormy, and nothing that happens can make them less so. Squaring this idea with the evolution of everything from fish to badgers to people from worm-like ancestors, if not actual worms, seems a bit difficult, but the nativists suggest that evolution confers people with "primitives" or core abilities in things like language and social interactions.

Those click-baiting headlines and our fascination with the notion that something like owning a dog could be related to one's DNA make it clear that this is much more than an academic squabble. The debate matters for many reasons. If, for instance we assume that intelligence is predetermined, we are less likely to think interventions, say in the form of social programs to improve learning in children, will be effective. On the other hand, if we assume everything can be altered by our actions, we may blame the victim, as is sometimes seen in the suggestion that those with cancer or other serious diseases could have avoided their plight by diet or exercise, or that people can just think their way out of depression. This doesn't mean we need to find the "correct" answer—as I will show throughout this book, such an answer doesn't exist—but that we need to be careful in how we think about the controversy.

Perhaps, ironically, we keep thinking we've resolved the debate, because we know—or at least we think we do—that traits as complex as intelligence or gender identity have to be affected by both genes *and* the environment. And yet, despite all the discoveries that show how interwoven learning is with what we call instinct, we keep either resurrecting the debate or optimistically declaring victory. Over thirty years ago, Daniel Koshland editorialized in the prestigious journal *Science* that "the debate on nature and nurture in regard to behavior is basically over. Both are involved."[6] Since AAAS publishes *Science*, someone clearly didn't get this memo when they wrote about the innate and learned behaviors of pets, but so be it. It is abundantly clear that this pronouncement was more than a little premature.

Psychologist David Lewkowicz, in his 2010 presidential address to the International Conference on Infant Studies, refers to the "biological implausibility of the nature-nurture dichotomy." A 2009 article by child development experts at the University of Iowa proclaimed, "Why We Should No Longer Abide the Nativist-Empiricist Debate." One of the many respondents to the article said that it isn't a debate, it's a dialogue, and a necessary one, which struck me as the kind of proposed middle ground that leaves no one happy. And yet Robert Plomin, whom I mentioned previously, will have none of that, and says instead, "In the Nature-Nurture War, Nature Wins."[7]

I could go on. But the truth is that the nature-nurture controversy has become what scientists call a zombie idea, one that, no matter how many times we think we have disproved it in the past, or reframed it, or even declared a victor, it springs back to life. In his blog Dynamic Ecology, Jeremy Fox calls these "ideas that should be dead, but aren't. Zombie ideas are the most important failures of science's self-correction mechanisms: they're big, widespread errors or misconceptions that aren't recognized as such."[8] His examples from the field of ecology are pretty much inside baseball for specialists, including things like whether species have more interactions in the tropics than in more temperate regions, but they underscore our inability to stop using an idea when its usefulness is long past its use-by date. Even if we believe that both nature and nurture are involved, which many of the people weighing in on the matter say, we then want to know which predominates, which one wins. Is it 50:50, 90:10? No one likes a tied game, it seems, much less one where you can't even tell who's playing for which team.

Why does this matter? Can't we just let the zombie wander the landscape, shedding DNA like rotten body parts and moaning about inheritance? (Jeremy Fox mused in 2012 that perhaps having the zombie apocalypse come to pass wouldn't necessarily be such a bad thing, at least from the perspective of the zombies.) But I am afraid we can't do that. We need to stop these fruitless debates about the inherent nature of sexism, or of genius, or of any one of a number of other topics that are central to our lives. A 2017 New York Times article titled "The Unexamined Brutality of the Male Libido"[9] argued that we are simply stuck with brutish men who can't or won't examine their destructive sexuality. Sorry, ladies, complain all you want—it's just immutable male nature. Like so many articles of this kind, it struck a nerve, with over 1,200 comments within hours of publication, some with furious condemnation, others expressing passionate agreement. Here's the problem: if people genuinely believe that men will always grow up with violent or dominating tendencies, then even if they see men behave otherwise, they will discount it, or at least they won't trust its permanence.

We also need to resolve our thinking about what is nature and what is nurture because we use our notions about what "comes naturally"—in other words, what is instinctive—in contradictory ways. For example, so-called women's work, whether it's done in the home or in "pink collar" jobs like nursing or teaching, can be seen as women just doing what comes naturally, using their maternal instincts or their innate ability to empathize to take care of other people's needs. If one then assumes that "coming naturally" means effortless, requiring no conscious thought or special skill, such a skill is not seen as "real" work—or in some cases, seen at all. But that's just as silly as suggesting that learned skills, like mountain climbing or solving crossword puzzles, are somehow more important or more difficult than ones that seem to arise from our genetic makeup. In the chapters that follow, I will show how it's the interplay between genes and environment that's important, and not one contributor or the other. We don't need to be either a blank slate proponent or a genetic determinist, and in fact those sides don't accurately represent the biology.

It's also important to point out that our obsession with changeable versus immutable is nowhere more obvious than when it comes to behavior. Sure, we wonder about how all of our characteristics got to be the way they are, but the knives are drawn when we discuss our aggression, our xenophobia, or our love of cats (or dogs). The headlines never blare, "Genetic Basis for Blood Cell Shape Revealed!" because we already assume that such characteristics depend on our genes; it's not as though we can will ourselves to have rounder lymphocytes. And yet all attributes, behavior or not, have contributions from both genes and the environment, in a complex way that defies a simple apportionment into percentages of each. Behavior seems so flexible, so ephemeral, that we can notice the way it seems to be inherited, as when twins share the same hobbies, but be perplexed that such fleeting attributes are passed on in the genes.

A Tangle of Blueprints

So if it's not nurture alone, and it's not nature alone, and it's also not some combined percentage of the two, where do our characteristics come from? The word that scientists use to describe how genes and environment work together to produce a trait (including a behavior) is *interaction*, which of course is used in common parlance as well. In biology, we use this word to mean that the effect of an organism's genes depends on the organism's environment, and the effect of an organism's environment depends on its genes. The philosopher of science Evelyn Fox Keller calls this the entanglement of genotype and environment, which also conveys the inextricable nature of the relationship between the two.[10]

We can illustrate the tangle if we review a seemingly simple case that is often considered the epitome of a single-gene defect in humans, phenylketonuria (PKU). PKU was the first condition widely used in screening infants, using the heel-prick blood samples taken from newborn babies. PKU is a simple disorder with devastating effects: babies with two copies of a defective gene cannot properly metabolize the amino acid phenylalanine, which builds up in their bloodstream and eventually leads to severe intellectual disabilities.

What could be a clearer example of a genetic cause? (Indeed, even genetics textbooks sometimes describe it as such, and historian of science Diane Paul has written extensively about the adoption of PKU as the poster child for genetic diseases.)[11] And yet this characterization is incorrect. It turns out that if these babies are given a special diet that makes up for the phenylalanine they cannot digest, they develop quite normally. Thus one could equally argue that the disease is completely environmental: it all depends on what they eat (or are given). Note that it is not just that both genes and the environment matter; their interaction is what matters. The outcome, whether the child grows up intellectually disabled or not, depends on which diet they receive only if they have the defective genes, or, equally, it depends on which genes they have, but only if they have an unrestricted diet. Even in this simple case, we cannot ascribe a greater cause to genes or environment.

You might object that, in the usual environment (an unrestricted diet), PKU is indeed an example of a genetic characteristic. But this objection actually proves my point: whether or not a trait can be considered "genetic" (nature) or "environmental" (nurture)—a very rare occurrence in any event—depends on the environment and the genetics, respectively. Of course, most characteristics, especially behavioral ones, are more complex in their genetics and the environmental variation less so, than in PKU.

Part of the problem may be that English doesn't have a good word for how the interaction—or entanglement—functions. We sometimes say that genes "underlie" or "underpin" behavior, as a way to distinguish their action from "determining" behavior, but these words still suggest that the genes are somehow fundamental, with the environment varying around them, which isn't the case either.

And speaking of terminology, the language used to discuss genes and how DNA is related to the way we are, whether that is physical or behavioral, is often fraught with confusion. The genome is sometimes said to provide a "blueprint," suggesting that all one needs to do to make a living creature is to follow the instructions. Even the website of the National Institutes of Health in the United States claims that "each genome contains all the information needed to build and maintain that organism."[12] Both ideas imply that the environment is secondary and the genes contain what's important, which is not true. Even Wikipedia lists the blueprint metaphor as one of the biggest misconceptions about genetics.

So what's wrong with saying the genome has instructions? For one thing, as Eric Turkheimer, a renowned behavioral geneticist who has studied human behavior for many years, says, "The relationship between blueprints and buildings is nothing like the relationship between genotypes and organisms. A blueprint of a house has a roof and walls and windows and a floor that have a direct one-to-one correspondence with the roof, walls, windows and floor in the house. There is nothing like this in DNA: there aren't head genes and feet genes, much less extraversion genes and divorce genes."[13] Another way to look at it is that one could reverse engineer a blueprint from a house—by looking at the way it is structured, one could reproduce, more or less, the drawings that directed its construction. But no one could look at an animal and say with any confidence what its genome is. Similar objections could be made to other common metaphors, like genes "coding for" characteristics, or behavior being "hardwired." The entanglement makes it impossible for such one-way processes to occur.

The Unhuman Solution

A way out of the conundrum is to think about human behavior in the context of other animals. Plato and Aristotle disagreed about where human knowledge comes from, but they did not think to question how a bird knows to build a nest, or how bees know to summon their colony to a rich source of nectar miles away. They probably thought the answer was obvious: animals operate by blind instinct, such as Plato's suggestion that humans are born with innate sources of understanding. Surely their senses are sufficient.

But this kind of human exceptionalism won't pass muster, particularly now that we understand that humans and other creatures all evolved from common ancestors. How could people need a special form of knowledge acquisition when all other animals happily go about their days without agonizing over the balance between nature and nurture? Only the most ardent human exceptionalist would cling to the notion that people—not chimpanzees, not dolphins, not even dogs—are the only species in which behavior springs from upbringing and upbringing alone, or the only species in which we need to consider the balance of upbringing and our biology. That doesn't mean that the alternative is a genetic determinism of the sort bemoaned by the *New York Times* writer regarding brutish men, that shrug of "what are you going to do, it's human nature." It means that the notion that genes "code for" or are a "blueprint for" behavior is simply wrong, whether you are talking about apes or people. I don't want to be disingenuous and claim that humans have no unique qualities, but it simply isn't reasonable for the nature-nurture controversy to apply only to humans, and to only human behavior at that.

This is not to say that we aren't fascinated by the similarities and differences between what animals learn (or don't) and what we learn. A 2021 article in *New Scientist* titled "Human-Like Intelligence in Animals Is Far More Common than We Thought" goes on to say, "Tests reveal that they don't merely act on instinct but can think flexibly, like us." Alternatively, coverage of a paper in *Nature*, one of the premier science journals, was headlined "Puppies Are Hardwired to Understand Us" and reported on a study that examined how well eight-week-old Labrador and golden retrievers responded to human cues like pointing. The title might be somewhat undermined by the statement: "Some puppies were more successful than others, but the researchers found that approximately 43% of that variation in performance was due to genetics," which suggests that the hardwiring has its limits. Commentary on the same paper in *Science*, *Nature*'s peer journal, stated, "If social intelligence is genetic, dogs should display it at a very young age. And there shouldn't be any learning required."[14]

But of course social intelligence, and everything else we (or puppies) do, requires input from both the genes and the environment (or learning). Behavior isn't genetic, whatever that means. It's that entanglement again, whether in dogs, humans, hippos, or hamsters. The question of whether nature or nurture is more important is impossible to answer. But we can ask a much more interesting question: How does behavior evolve? How do the characteristics of everyone, human and animal, come to be? This book explores that question, by examining how genes and the environment act in concert to produce behaviors as different as fetching in dogs and chirping in crickets.

Recall the example of brutish men, but cast it in a new light. Men are not genetically programmed to cheat on their mates or sexually harass their coworkers, but neither are they programmed to be caring and cooperative. Genes don't work that way. By the same token, the social milieu or the patriarchy are not the only factors that affect their cheating (or caring). The effect of genes on behavior depends on the organism's environment and vice versa. It's not just that nature and nurture both contribute, it's that you can't meaningfully consider the action of one without the other.

In the chapters that follow, I look at how behaviors in a wide range of animals evolve, and how we think about the role of genes and the environment when it comes to both animal and human behavior. Chapter 1 lays the groundwork by asking a question that turns out to be much more difficult to answer than it might first appear: What is behavior? This question has a companion: Is behavior special? In other words, is behavior somehow different from physical characteristics like the shape of a nose or how much fur your cat has? Chapter 2 examines what we know about the way behavior evolves, making use of Iranian snakes and skydancing humming-birds. It also takes a good hard look at the idea that humans, or any other mammals, have a "lizard brain" underlying our more sophisticated actions. Chapter 3 details how much genes can tell us about behavior, starting with some work on fruit fly courtship by one of the rare early women evolutionary biologists and showing what it really means for a behavior to be heritable.

In chapter 4, I use the domestication of our most familiar animal companion, the dog, to see just how much behavior can change with a few thousand years of cooperation, and cast some doubt on the notion of a domestication syndrome. Chapter 5 continues the domestication theme with cats but also including some other species, like guinea pigs, which turn out to have attained their iconic status as experimental subjects because of the anti-vivisectionist stance of George Bernard Shaw. Chapter 6 takes a look at mental disorders in animals, with crayfish that are anxious and dogs with a form of obsessive-compulsive disorder. Chapters 7 and 8 survey both animals that we think are clever, like ravens, and those we have often dismissed as being mere automatons, like bees and crabs, asking just what we mean by birdbrained and whether intelligence is all about the brain. Oh, and we'll also consider the assfish, and no, I am not making that name up. In chapter 9, I will explore the evolution of language, or what seems like it, in animals, trying to determine whether the Rubicon of human speech is as uncrossable as some would have you believe. Chapter 10 visits the thorny topic of sex and gender, with another look at whether men, or male animals in general, are all that inherently brutal. And finally, because the COVID-19 pandemic made us think about disease in our lives in a wholly new way, chapter 11 is about how behavior that fights disease—from birds fumigating their nests with cigarette butts to ants performing battlefield triage to chimps chewing leaves that rid their guts of worms—can evolve. In the end, I hope that understanding how behavior evolves helps us see both other animals and ourselves better, and that the battle between nature and nurture is not worth fighting.

Narwhals and the Dead Man:
Why Is Behavior So Hard to Define?

sometimes show three videos to students in animal behavior classes. First is a male peacock spider displaying on a log. These Australian natives are smaller than your pinky fingernail, but they engage in a vigorous semaphore-like dance, with legs flailing rhythmically and the iridescent multicolored abdomen flung upward (one YouTube version memorably sets the display to the Village People song "YMCA," which turns out to be remarkably suitable).

The second video is of white blood cells consuming a pathogen. It is less colorful than the spider, but it shows the cells moving through liquid and then finding and surrounding clumps of bacteria, eventually engulfing them without a trace left behind.

The final one is a Venus flytrap closing on a fly. Accompanied by cheesy suspenseful music, we see the insect hover near the open trap and then alight inside. After a second or two, the flaps of the plant's trap begin to close, its fingerlike projections interlacing around the fly. We can see the fly struggling and hear its feeble buzz as the trap tightens, but there is no escape, and the fly is left to be digested by the plant.

My question to the class is this: Which of these do they consider to be behavior? As you might imagine, everyone agrees on the spider. Perhaps half are on board with the cellular action of the white blood cells, and a small but passionate minority defends the plant as having behavior on a par with that of animals. I always let the discussion go on for a while before I weigh in, partly to encourage them to think rather than wait for the "right" answer, and partly because the difficulty of defining behavior, paradoxically, makes it easier to see how it evolves. For the record, I am with the majority—spiders behave for sure, but cells don't unless you take such a broad perspective that behavior itself becomes meaningless. If the cells behave, then we can even talk about atomic behavior or the behavior of the planets and stars, which gets us pretty much nowhere.

Consciousness Flowering

Those students on Team Flytrap, who see behavior in a plant that seems like it has a kind of chlorophyll-fueled attitude, are not alone. Flytraps even raise the question of whether they not only behave, but whether they can count. This is not as far-fetched as it may seem. When a hapless insect lands on the "trap" (actually a pair of leaves) of the plant, it triggers the leaves to snap together when it contacts hairs on the inside of the trap. After the trap closes, the insect struggles to escape. But the trap doesn't close upon the first contact with the hairs—presumably a jittery plant that clamped its leaves shut every time a stray bit of trash or even a strong gust of wind contacted the hairs would waste its energy, and it would never manage to capture a decent meal. Instead, the plant's hairs start the closing process only when they receive two signals within about twenty seconds. Furthermore, once the trap is shut the process of digesting the captured insect only begins after three separate signals. Not one, not two, but three. If this doesn't sound like much of an accomplishment, remember that children don't develop the ability to count until they are somewhere between fifteen and eighteen months old.

Whether or not you admit to the counting, does the response

of the flytrap to the insect qualify as behavior? To answer that, we need to know what behavior is, whether in people or animals. This turns out to be a surprisingly difficult question to answer, and one of interest to scholars from several fields, from psychology and biology to philosophy. As recently as 2019, psychologists were still wringing their hands about not being able to define the activity at the core of their discipline's existence. Psychology professor Raymond Bergner from Illinois State University rather peevishly observed that psychologists who study behavior without defining it "are in effect like a biologist who, in introducing a treatise on 'vertebrates,' informs readers at the outset that he or she cannot tell us what a vertebrate is."[1] Leaving aside that, as it happens, vertebrates are not always easily distinguished from some of their more oddball relatives (google "tunicate" if you don't believe me), he has a point.

Psychologist Gregg Henriques from James Madison University, along with Joseph Michalski at Kings University College in Canada, start a lengthy 2020 paper with the rather gloomy statement: "Even though the concept of behavior is central to modern psychology, there is no consensus regarding what the term behavior means."[2] I usually tell my students that behavior is anything an animal or human does, but I recognize some inherent shortcomings in that definition.

In 2009, Daniel Levitis, a biologist now at the University of Wisconsin-Madison who has studied topics ranging from aging in barnacles to longevity of fungi, together with colleagues published the results of a survey of members of three scientific societies with links to the study of behavior.[3] As with my videos, the survey asked respondents to say whether a given statement did or did not describe a behavior, including "a person's heart beats harder after a nightmare," "a spider builds a web," and "a rat has a dislike for salty food." They got 174 responses, and found, in the words of Natalie Angier from the New York Times, who covered the debate, "an impressive lack of accord."[4] The scientists not only didn't agree with each other, they also weren't always consistent in their individual responses to the different prompts.

The authors saw this as a matter for concern, and they tried to come up with a definition they could live with and thought would be useful: "The internally coordinated responses of whole living organisms to internal and/or external stimuli." So the white blood cells are out, because they aren't whole organisms, but parts of one. Plants, though, including of course the Venus flytrap, are fair game.

Whether plants behave turns out to be part of a larger issue, namely whether plants are conscious, the subject of a rather testy debate in the last decade or so. In 2006, a paper in the journal *Trends in Plant Science* proclaimed the emergence of a field called plant neurobiology, which at the very least puts plants right up there with the frogs you stabbed in high school biology class when you examined their reflexes.[5] Studies published in that journal and elsewhere claim that plants communicate, that they have true memory, and that they perform tasks with intention. A flower that changes its response to pollination depending on how recently an insect has visited it (or a scientist imitating an insect) was designated as "savvy" in the *New York Times*. A cactus that absorbs ultrasonic waves via the structure of the hairs surrounding its flowers was said to "entice" the bats that pollinate it. Journalists rejoiced, cute headlines were written, and thought pieces about plant intelligence appeared in *Forbes* and the *New Yorker*.[6]

Not everyone, however, was impressed. A group of scientists from the United States and Germany wrote a paper titled "Plants Neither Possess nor Require Consciousness" in which they pointed out that the plant consciousness proponents simply glossed over the complexity of the brains and nervous systems of even relatively simple animals such as worms (not that this necessarily means that these creatures have consciousness, just that the building blocks are certainly not present in plants). The lead author, Lincoln Taiz, acknowledged that if people believe plants are conscious, they might be more likely to support environmental causes that preserve all living things, but that this worthy goal doesn't justify drawing unwarranted conclusions.[7]

Taiz and his coauthors note, too, that people have been anthropomorphizing plants for many years, with a particular emphasis on plant sex lives and pollination. At least as early as the eighteenth century, biologists described a passion in plants that, as Taiz's paper puts it, "could have been written by D. H. Lawrence," including "excited organs, which seem to think only of satisfying their own violent desires." Carolus Linnaeus, the Swedish founder of the scientific classification of living things, had his own ideas, often discussing the "bridal bed" of plants and the interactions between pollen and ovule as "marriage," so that a plant with nine stamens and one pistil was "Nine men in the same bride's chamber, with one woman."[8] Whether this made people of the era more sympathetic to plant conservation, or even to the idea that plants have agency or consciousness, is hard to say. It might have made them more alarmed by botanists.

My own stance is skeptical, along with Taiz and his colleagues. Those time-lapse videos showing leaves unfurling or tendrils extending during growth certainly look like intentional behavior, similar to a cat seeking out a sunny spot for a nap, but that's mostly because the sped-up frames fool our eyes by mimicking an animal's movements. The internet is rife with videos of inanimate objects that seem like they are out to get us. We are also living in the age of Roomba, but that doesn't mean that the science fiction fear about the rise of the robots has come to pass. And as for lustful flowers, or innocent bridal plants, that's plain creepy. Just because we can anthropomorphize something doesn't mean we understand it, and at least in the case of the botanists turned erotica writers, it says more about the authors than their subject does.

Ain't Misbehaving?

Leaving aside consciousness, whether in plants or anything else, asking whether something behaves, and what behavior means, is still an open issue. Part of the problem is that behavior is studied both by psychologists mainly interested in human behavior and by biologists like me, who are happy to include humans but whose main focus is on animals. Can we come up with a workable definition that satisfies both camps?

We aren't there yet, but people are trying. In Henriques and Michalski's paper, the universe is divided into a tree-like hierarchy, with matter (associated with the physical sciences) at the base, life (biological sciences) next, the mind (psychological sciences) after that, and culture (social sciences) spreading at the top. (Somewhat to my disappointment, the paper did not contain a single "mind over matter" pun.) The system thus includes behavior in "1) material objects (e.g., hydrogen atoms); 2) organisms (e.g., bacterial cells); 3) animals (e.g., dogs); and 4) people."[9] The authors spend quite a bit of time explaining how this hierarchy eliminates the problem of psychologists never grappling with what bacteria do, despite Levitis's definition keeping microbes firmly in the purview of behavioral biologists.

I wondered quite a bit about that list. For one thing, animals are organisms, as are plants and fungi, along with the microbes they list. The term *organism* just means living thing, and even if we save the discussion about what constitutes life for another day, it doesn't help much to split up creatures great from those that are small. Moreover, who feels compelled to give dogs as an example of an animal? Don't we all know that dogs are animals? And what about sponges, or other living things that aren't easily classified as animals? The problem is that any time you start creating these boundaries, whether between kinds of objects in the universe or between behavior and other characteristics, fuzziness follows. That doesn't mean there aren't any distinctions, but that you had better know why you are creating the definitions in the first place.

In a way, it becomes easier to create such distinctions the less you know about the diversity of living things and their activities. I am willing to bet that few psychologists think much about slime molds, for example. These organisms—it's necessary to use that term, because they aren't plants or animals, and they used to be considered fungi but are no longer—live quiet blameless lives on forest floors and lawns, places where they consume bacteria, yeasts, and fungi. One type has the misfortune to have been dubbed the "dog vomit slime mold," which seems unnecessarily harsh to me. Nevertheless, the group is fascinating for several reasons, not least of which is their ability to seem as though they can solve problems and predict the future even though they look like blobs of glup, as James Thurber described the fearsome Todal in his story *The 13 Clocks*. People who study them refer to slime molds as "nonneuronal," which sounds like a politically correct way to say brainless without the pejorative associations.

Be that as it may, slime molds will sometimes forage as a lone cell and sometimes form groups of cells that all have the same genetic makeup. It would be as if you mostly existed without your tissues and organs, but then acquired a liver or kidney as the need arose. These groups of cells will ooze through their environment, making tubes that can go in one direction or another. The basis for that decision has much in common with the way that we and other "neuronal" beings figure out a problem.

Chris Reid and colleagues from Macquarie University in Australia offered slime molds a maze with two arms, so that the creature had the option of staying and consuming food in one or switching to see if there was something better in the other. This problem is called the two-armed bandit, named after casino slot machines. The arms in the slime mold experiment had different amounts of food, and the slime molds, like most creatures, prefer larger amounts more than smaller ones. The slime mold could initially explore both arms by extending itself into each. Then, the test: since it could choose the better arm, it should stop exploring and consume only at the better option. Because the slime molds extend over the area where they feed, it is a simple matter to determine which arm is preferred by weighing the amount at each location.[10]

It turns out that slime molds, like fish and birds (at least the few kinds of fish and birds that have been tested), solve the problem in a reasonably optimal way, settling down at a high-quality food source more often than not. It would be hard not to call what they do behavior, and the authors of the study go even further, saying: "These similarities raise the compelling notion that deep principles of decision-making, problem-solving and information processing are shared by most, if not all, biological systems."[11] I have to concur, but then I have always liked slime molds. Perhaps we should rethink that dog vomit moniker.

Testing the Dead Men

A different group of psychologists has a very practical reason for defining behavior. Behavior analysts, as they are termed, follow in the footsteps of the American psychologist B. F. Skinner, famous for his reliance on reward and reinforcement to shape behavior. The field is experiencing a resurgence because its methods are employed in therapies for children on the autism spectrum, among other things. Behavior analysts focus on changing behavior by changing the people's or animals' relationship to something affecting them in their environment.

Before one can shape behavior, however, whether that is encouraging children to brush their teeth or helping someone overcome substance abuse, behavior itself needs to be defined, and here behavior analysts have one of the most memorable definitions I have ever run across. Drawing from the work of psychologist Ogden Lindsay, behavior analysts employ the Dead Man Test (DMT). It simply states that if a dead man can do it, it isn't behavior; but if he can't, then it is. One behavior analysis clinician includes the following criteria of behavior on her website:[12]

- Anything a person says or does
- Involves movement and has an impact on the environment

- Is influenced by environmental events
- Can be observed, described, and recorded
- Needs to pass the dead man test (teddy bear test)
- If a dead man or teddy bear can do it, then it IS NOT behavior!!!!

Presumably the last two entries soften things a bit, with a teddy bear being nicer to envision than a corpse.

That may seem clear cut, and it is presumably obvious that, as behavior analysts Thomas Critchfield and Elva Shue point out in a 2018 paper, "Behavior is absent in vitality-challenged individuals." Nevertheless, defining behavior gets us into some pretty muddy philosophical waters, impinging on what we think about consciousness, cognition, even the definition of life itself. And indeed, philosophers have often tackled the question of what constitutes any of these, and whether behavior or cognition should be confined to humans, or animals, or should include plants and one-celled creatures. A group of Dutch philosophers suggests that our definition of cognition, at least, could become more inclusive, and proposes bringing into the fold the bacteria *Escherichia coli*.[13] This bacterium, at least in some forms, can cause serious gastrointestinal disorders if it contaminates produce or meat. It may or may not be a consolation to know that the same bacteria can respond to chemical changes in their environment by "running" or "tumbling" (these are terms used by microbiologists to characterize ways that the cells propel themselves), and these movements are accepted, at least by the philosophers, as qualifications for what they refer to as "minimal cognition." As with the white blood cells I use in my examples with students in the introduction, I'm not so sure we get much out of putting microbes into the cognitive tent. I'm at least willing to entertain the idea that they behave, but cognition seems a bridge too far.

What Would Nemo Do?

The arguments in my classes over definitions of behavior are often useful in getting students to be precise and think about what they mean when they speak. But I have a somewhat different point to make about the murkiness of distinguishing behavior from other characteristics, or of deciding who behaves and who doesn't. The point is that if we can't draw a hard and fast line separating behavior from physical traits, then the same rules apply to both, and behavior evolves the same way that leg length or other physical characteristics do.

This is an important conclusion, because it means that we can't invoke "culture" as a get-out-of-evolution-free card. What I mean by that is that it's easy to talk about the evolution of physical characteristics such as a four-chambered heart, and the degree to which such characteristics are influenced by genes. Ancestral reptiles that could shunt oxygenated blood from the lungs to the body more efficiently had more babies than those that could not. What about behavioral characteristics, such as intelligence or talent for playing the piano? They seem as though they must be more complicated because they are so heavily influenced by culture, the environment, or our upbringing. Hence, the argument goes, we have to set aside behavior, particularly human behavior, as not evolving the same way as physical characteristics, and we also shouldn't think about behavior evolving through selection on whichever variant leaves the most offspring. It's just too ephemeral.

But if, as I hope I just showed, behavior isn't different from physical properties, then trying to create separate rules for the evolution of behavior and appearance or physiology is an exercise in futility. If you make a sharp distinction between behavior and physiology, then Pavlov's dogs don't behave when they salivate in response to the bell.

The similarity between behavior and other traits doesn't mean that genes dictate behavior—far from it. Instead, it means that all of those other characteristics, like height, are subject to influence from the environment as well, and both genes and the outside world act in concert to shape every trait, whether behavioral or not. They all evolve in the same entangled way.

An illustration of the blurred line between behavior and other characteristics can be seen in clownfish, those cheerfully striped inhabitants of sea anemones featured in the film *Finding Nemo*. Like many movies, *Finding Nemo* got some things wrong on the biology of its star. To start with, young clownfish don't live with their dads by themselves; instead, they are found in groups with one male and female and up to four nonbreeding individuals.[14] The breeding female is the largest fish, the male second, followed by the others in a neatly descending line of body sizes, like the von Trapp family if they lived underwater. If the female dies, the male gets bigger and also changes his sex, undergoing a complex set of physiological alterations to its internal organs. The rest of the group then adjusts accordingly, with the largest nonbreeder becoming male and everyone else ratcheting up size in strict order. Since only the two biggest fish get to reproduce, size really matters.

Peter Buston is a marine biologist at Boston University who has been studying clownfish in Madang Lagoon, Papua New Guinea, since he was a doctoral student. Originally from the UK, he jokes that "only an Englishman could have uncovered the perfect queue in a group of social fish."[15] But it's more than a convention that keeps the peace among Nemo and his friends; the way that clownfish adjust their sizes underscores how flimsy the distinction between behavior and other traits can be. Buston and his colleagues saw that the fish could regulate their growth depending on the social surroundings—if you remove a second-place fish, the one in third place grows to fill the vacancy, precisely attaining the size of the individual that was taken.

What would happen if two fish of the same size both got to an empty anemone at the same time? Buston re-created such a situation in his lab, putting two clownfish into a tank and comparing their sizes after two to three weeks to a solitary fish kept under the same conditions. Amazingly, the paired fish grew much faster than the singletons, even though the amount of food they were given, and every other characteristic the scientists could manage, were kept the same. The clownfish can gauge how much they need to increase their size in the group to get to about 80 percent of the weight of the fish ahead of them in line—exactly how they do so is a mystery. This amount allows for rapid response to achieve the next level, but doesn't pose a threat to the dominant individual. This is called competitive growth, and it shows just how tuned to the social environment a seemingly basic characteristic can be. According to Buston, "The most interesting thing clownfish do is grow."[16]

The moral of the story (aside from chiding Pixar that Nemo really should have hung around his dad, who really should have become his mom, except that baby clownfish don't stay with their parents . . . oh, never mind) is that growth is just as powerful an indicator of identity as fighting ability or any other "real" behavior. Competitive growth is also seen in another animal that is in a beloved children's movie, the meerkat featured in The Lion King. In this case, the investigators, led by Tim Clutton-Brock at the University of Cambridge in the UK, took pairs of littermates and fed one lucky member a hard-boiled egg a day for three months in addition to their other food, which led to accelerated growth. The other member increased its own growth rate in response, much more than meerkats that were not given the supplement but were housed separately.[17]

How do the organs, tissues, and the brain work together with the genes to produce an individual animal that continues to respond to the world with behavior? To answer this question, consider a goat that was born with two legs, described in 1942 by a Dutch scientist named E. J. Slijper.[18] This goat was born with a defect—whether congenital or environmentally induced is unknown—in which the forelimbs did not develop as usual, and so it could not walk on all four legs. Instead, the goat learned to walk and even run, using just its hind legs. After the animal died, Slijper performed an extensive autopsy on the goat. He noted striking differences between this goat and the more normal four-legged variety; the bipedal goat had much enlarged hind limbs, a curved spine, and a very large neck.

The goat, of course, remained a goat, and it would not have passed its bipedal abilities on to its kids, if it had been able to have any. Why, then, does it matter for our thinking about behavior and how it evolves? It matters because the goat shows how that entanglement between genes and the environment allows behavior—and physical characteristics—to evolve.

Slijper noted a surprising parallel between the goat's musculature and that of a kangaroo, which routinely relies on its hind legs to move. Similarly profound changes in body structure can be seen in humans who use their arms to compensate for the loss of legs, or in apes and other primates made to walk on their hind legs without using their arms for support. Using the muscles differently makes the bones develop into different shapes and sizes. Those alterations would ordinarily be expected to require generations of evolution. But in the goat's case, a physical change—the defect—led to a completely novel behavior—bipedalism—which in turn caused pronounced physical changes—bigger bones, differently shaped muscles—that then permitted the animal to show a completely new and apparently functional form—a goat with two legs.

Importantly, the goat became bipedal without needing a new gene or genes that made that possible. Not having legs didn't mean that another gene making bigger muscles somehow came into being. Instead, the same genes did different things in response to changes in the environment, in this case the environment of the goat's own movements. What genes do isn't a one-and-done event; it is an ongoing process, and one in which the environment affects the genes, the animal changes its behavior, that behavior then affects which genes are used, and so on. Genes aren't simple computer programs—they change their directions in response to how they are used. The reason that the two-legged goat could adapt to its circumstances had to do with the way genes are signaled to turn on and off. The goat had the same molecules and chemicals as any other goat. It didn't lack a "gene for a front leg." Instead, the genes that would ordinarily have directed enzymes to make muscles in the front leg activated as they normally would, but the muscles attached to different tissues because there was no bone available.

Slijper's goat died without leaving a mark in the goat world. But say it had not died. If it had gone on to survive in an environment where being bipedal was an advantage, changing the place where it lived, and if it had then been subject to selection that made it survive and reproduce better, maybe we would have seen the evolution of two-legged goats. The trajectory of evolution can change because behavior, in this case the ability to walk upright, got entangled with the genes.

I have been operating all along under the assumption that behavior does indeed evolve, but not all scientists agree. Notably, Peter Klopfer, an eminent, now-retired Duke University professor who studied mother-infant behavior and aggression in several kinds of mammals, especially lemurs, does not agree. Klopfer is not a creationist, but he declared in several publications that behavior doesn't evolve. Part of his criticism arose from his reluctance to consider behavior in the same category as physical characteristics in the way that I have advocated here; he was particularly derisive about the idea that behavior is an entity with neat boundaries. Or, as he put it in a review of a book about the ethology of mammals, "We read of 'aggression' accumulating and needing discharge, as if it were a fluid liable to seep through cracks in the cranium. I believe we 'contain' aggression about as much as a radio 'contains' the music we hear issuing from it."[19]

Klopfer's reservations stemmed from the lack of a connection between a specific gene or gene product and the changeable, messy actions that we call behavior. After all, one can never find a gene that will be linked forever and always to a mother monkey's embrace of her infant, or even a bee's preference for a yellow flower over a blue one. All behaviors are profoundly influenced by the environment. He also expressed skepticism about using behavior in evolutionarily related groups—a wing display in one kind of duck that is seen in a modified form in another—the way we look at the fossil records to understand how whales arose from four-legged ancestors. I will address the latter phenomenon, called homology, in the next chapter. As for his former concern, it is becoming increasingly clear that the environment influences all characteristics, including physical ones like the shape of one's organs, not just behavioral ones, and that genes and the environment interact for all of them as well. Your liver is the size that it is because of your genes interacting with your diet, for example. Thus, if we are in for an evolutionary penny, we have to be in for a behavioral pound.

Be Still, My Narwhal Heart

The blurring of the distinction between brain, behavior, and body is of more than intellectual interest. It can predict how well animals can be expected to survive in new environments, a situation that humans have made much more common over the last several decades. When behavior and physiology collide, it can have dire consequences for the animal. Recent work on an unlikely poster child for human interference with behavior reveals those consequences.

Terrie Williams, a marine scientist at the University of California, Santa Cruz, has spent decades studying marine mammals, including how their bodies are able to dive to amazing depths to find food. Williams recently examined narwhals, the single-tusked creatures that live in the frigid pack ice of the Arctic.[20] (Yes, she knows they are called the unicorns of the sea, but she wishes people would get over it—at a conference we both attended, she pointed out that they are not all that exotic, and we can just think of them like any other whale. Also, males do not fight with their tusks.)

Narwhals can dive 1,500 meters, or nearly a mile, as they hunt squid and fish. When they descend, they have specialized mechanisms that allow them to exert themselves without using too much oxygen, since they hold their breath during a dive. Most notably, they slow their heart rates way down, to three to six beats per minute. It's hard to wrap your head around that number. Williams's recording of a narwhal at the surface sounds more or less like the regular seventy-something beats per minute we are used to—lub-*dup*, lub-*dup*, lub-*dup*. Her recording of the diving narwhal is obtained using an ingenious device that attaches harmlessly to the whale's skin, rather like a tiny portable electrocardiogram machine. Lub-*dup*. Pause. About thirty seconds passes. Then, finally, just as most of us in the audience had started to wonder if something was wrong with the audio in her presentation, it came: lub-*dup*. No Olympic athlete could ever match that funereal heartbeat.

This extraordinary adaptation, however, can get narwhals into trouble. Narwhals are mammals, and mammals have a physiological reaction to stress or fear that takes one of two forms: freezing, or the fight or flight response. Either one can deter a predator, and the mode employed by a given species is generally the one that is most useful to it. Many rodents, for example, will freeze when startled, which in natural circumstances allows them to blend in with the background and avoid detection by either a searching fox or a human hunting for pests in the garage. The responses have different physiological components, as you might imagine, since running for your life requires different muscles and nerves than staying completely still.

Under ordinary circumstances, narwhals are rarely subjected to predators other than occasional orcas or Inuit hunters, a benefit of living in one of the remotest areas on the planet. They are able to avoid such threats by swimming very slowly under the ice or moving to the shallows where they cannot be pursued.

But their solitude is increasingly encroached upon by ships, particularly those used in naval and other exercises. And when a narwhal is stressed by noise or other disturbances associated with humans and their ships, it has what Williams calls a "paradoxical response"—its muscles speed up for fleeing and its metabolic rate ramps up, as it would for a fright reaction, but its heart rate continues to slow as it does during a dive. These contradictory effects can have disastrous consequences. In mammals, the combination of a lowered heart rate and a sudden flight reaction can lead to cardiac arrest. Williams is adamant that the huge increase in shipping traffic around the narwhals' range poses a serious threat to their existence.

The narwhals' plight, alarming as it is, also illustrates the exquisite entanglement of brain, behavior, and body. In this case, the mixed signals ("Flee! Something dangerous approaches!" versus "Slow down! You are in very deep water and need to save your oxygen!") are a potential calamity, but they also show how impossible it would be to separate the behavior of swimming, or the sense of fear, from the physiology and anatomy that produces them.

A Cultural Interlude

Behavior, then, evolves the same way that physical characteristics do, via individuals with traits that enable them to survive and reproduce, which means they are better at passing on their genes than not-as-adept individuals. As a result, in the textbook definition of evolution, the genetic composition of a population changes accordingly. But the genes aren't all that matters. Both behavior and physical appearance are, as I have just been arguing, the result of both genes and the environment, inextricably entangled. And in thinking about human behavior, people refer not just to simple environmental influences like the climate or diet, but also to that amorphous thing called culture. No one would argue that culture affects how we behave; we talk endlessly about cultural influences on everything from our clothing and food preferences to our values. But culture is just another way of referring to the environment, and this is what I mean by culture not providing a get-out-of-evolution-free card. Saying that intelligence is affected by culture doesn't mean that it isn't influenced by genes, because everything is influenced by genes.

What, then, does the effect of culture have on how we see the evolution of behavior? Much ink has been spilled arguing about the way in which cultural practices like fashion, or ways of making pottery, or even religious beliefs, can change through time in a manner analogous to the way that fingers became wings or four-legged creatures turned into whales. Culture, in this view, has its own evolution. In his 1976 book *The Selfish Gene*, Richard Dawkins famously coined the term "meme" to refer to the unit that allows propagation of such human products, setting out to create a deliberate analogy with the genes he dubbed "replicators."[21] Now, of course, memes have become associated with images or ideas transmitted online, and most of my students are shocked to discover that the idea that led to Grumpy Cat goes back nearly half a century.

The notion that cultural evolution is its own process has gained traction among a number of scholars, with some drawing elaborate analogies to genetic evolution. Culture is sometimes used as a catchall for all the influences on organisms that cannot possibly have come from genes. So, for example, in an argument over whether there is a genetic influence on intelligence, one could say that a person might be viewed as smart in one culture and not so smart in another, simply because of different local emphasis on what skills are important. While undeniably true, that observation still doesn't negate the role of the genes, any more than acknowledging that people fed different diets may end up at different heights.

The other reason that claiming "human behavior is the result of culture" doesn't allow us to set behavior aside is that animals have culture too. If you define culture as shared traditions passed among members of a group, then lots of practices among animals qualify. For example, many birds sing slightly different songs in different places; the variants are called dialects, and like their human counterpart, they are passed along within the community, so that a White-crowned Sparrow from San Francisco has an "accent," as it were, that distinguishes it from one that grew up in Southern California. You can tell it's a White-crowned Sparrow with no difficulty, but its voice betrays its origin. The differences between songs of one place or another aren't inherited through genes, because a sparrow from one place reared with sparrows from another will sing its adopted community's song, just like a child from the United States raised in Korea will speak Korean.

Chimpanzees have long been said to have culture, and a study published in 2020 confirmed that the apes' methods for collecting termites, a highly desirable food, depended on the group to which the chimps belonged.[22] All chimps use sticks to fish for the termites in their burrows, but camera traps showed that each community had a slightly different and consistent technique: leaning on the elbows, lying on one's side, shaking or not shaking the stick, and so forth. It's not that one way is better than the other; it's just that an individual chimp will do what its companions do. The differences among groups are due to the happenstance of where a chimpanzee grows up. Other examples of culturally transmitted behaviors include the opening of milk bottles by a chickadee-like bird called a tit, the making of tools by New Caledonian Crows, and the wearing of sponges on dolphins' snouts to forage for fish.

No one would argue that opening a milk bottle is on a par with building the Sistine Chapel, or that cultural variation among humans is exactly like that of chimpanzees or birds. But the point is that both humans and nonhumans share the same mix of genetic and environmental influences on what they do. Culture is just another way of referring to the environment.

Are Humans Exceptional?

Even those who are happy to accept behavior evolving in crows or damselflies may balk at the same process happening in humans. Perhaps it is because we guard our culture so fiercely. Nevertheless, we have wondered about our place in the world, and our relationship with other animals, for millennia, and maybe longer than that.

Even modern scholars have made impassioned arguments for that key characteristic that makes us not like all the rest. Is it language? Tool use? Warfare? All of those?

Even specialists in animal behavior have had trouble deciding where to place humans in relation to other animals. The renowned animal behaviorist Niko Tinbergen, broadly acknowledged as the father of ethology—the study of animal behavior—certainly included humans in his field of study, flatly stating in his 1951 book *The Study of Instinct* that "man is an animal."[23] The ethologists at the time were fascinated by the stereotyped patterns of behavior they saw in wild animals; a goose would retrieve an egg moved from her nest using the exact same set of movements every time, which the ethologists thought indicated an instinctive—what we would now call genetic—behavior. Tinbergen was enthused about the idea that human behaviors could be explained using the same principles he had developed for studying wasps and gulls. Anthropologists during the 1960s were more skeptical, cautioning that human behavior, at least in their eyes, was much more influenced by the environment than the genes—one of the many examples of an assumption of the false dichotomy between the two. Tinbergen and other ethologists saw humans as animals with a "layer" of culture superimposed upon it. More broadly, that view was expressed by anthropologists; for instance, Clifford Geertz, who in 1966 referred to the "stratigraphic" nature of humans, a biological core surrounded by layers from psychology and culture.[24]

Aside from making me think of people as if we were tantamount to ice cream bars, that perspective strikes me as nothing more than special pleading. Other animals have psychology, if by that you mean behavior that is influenced by internal processes. And other animals certainly have culture, as I outlined previously. It's easy enough to make a list of things people do that animals don't: pay taxes, use flush toilets, write poetry. One could certainly also make a list of things animals do that humans can't: echolocate, metamorphose into a different life stage, regrow a limb. But it's not clear what we gain by finding one characteristic that seems qualitatively different from the rest.

Besides, each time we think we've figured it out, someone finds an animal doing something that we thought was uniquely human. As I will detail in later chapters, chimps can experience what looks like empathy, birds can not only use but also manufacture tools, and octopuses can solve complex spatial problems. Primatologist Frans de Waal notes that "cognitive science has blown big drafty holes in the wall supposedly separating us from the rest of nature."[25] It's true that the human and animal versions of the supposedly unique character- istics differ. Author Jamie Milton Freestone scoffed that if you define tool use broadly enough, it "can include both the sticks trimmed by crows to catch insects and laparoscopic surgery; the octopus's use of a coconut shell as a shelter and a 3D printer; the rocks otters use to smash open abalone shells and the Large Hadron Collider."[26] But where and how do you draw the line? Is the use of a coconut shell as a shelter qualitatively different from cowering in a cave? Ants create elegant and complex nests out of earth, akin to adobe buildings in the southwestern deserts of North America, so do we group them with the octopus, or with our early hominin ancestors?

My favorite example of a distinguishing characteristic of humans comes from a non-majors biology course in which I was a gradu- ate teaching assistant at the University of Michigan. To enliven the material in those pre-internet days, the professor had us show films during the discussion sections of fifteen to twenty students. One such offering was an episode of the BBC documentary *The Ascent of Man*, narrated by Jacob Bronowski. In tracing the evolu- tion of humans, Bronowski mused on what set us apart from other animals. I sat at the back of the room and listened with half an ear, making sure that the ancient film projector was still functioning, and assuming he would name tools, or the aforementioned speech, or maybe complex societies. Instead, his mellifluous Germanic voice paused. "And that is . . . frrrrontal copulation!"[27] He declared that no other animal had the ability to mate face to face, and that this enabled humans to fully develop our social aptitudes, though why that had to happen via intercourse rather than some other means was not made clear to me.

The class suddenly became alert, though no one's eyes moved away from their desk. After the film was over, the students were silent. Then one young woman raised her hand, looking skeptical. I called on her. "How the hell does he know whether we are the only species that has sex like that?"

A reasonable question, to be sure. It turns out that we are not, in fact, the only ones; others include dolphins and the ever-popular bonobos. And I'm willing to bet that some invertebrate, somewhere, mates face to face as well. But even if we found a whole tribe of worms that engages in missionary-position sex, it's not clear that would say anything about the uniqueness of the human condition.

Just as behavior itself is at least sometimes part of a continuum with physiology, what humans do has to be on a continuum with what other animals do. This doesn't mean that all living creatures are arranged on a ladder of progress like the ancient Greeks believed, with worms at the base and humans at the top just below angels, but that characteristics are shared in ways that are messy and complex. Once we see a continuum between characteristics of humans and those of other animals, it makes it easier to understand how behavior evolves in either one.

Bergner, the psychologist who asserted the need to define behavior, goes so far as to say that "the claim that the science of psychology will one day be superseded by that of biology emerges as a promissory note that clearly, at present, and arguably in principle, cannot be paid off."[28] In other words, something spooky is going on here, something that is more than the nervous system and its interaction with the senses and muscles. Human behavior, he thinks, has a unique quality that just can't be explained by the same rules that govern animals.

He may mean consciousness, a whole other can of metaphysical worms, but even if so, the idea that psychology contains a mysterious elixir that we cannot explain means that behavior would differ from anatomy or physiology, both of which are included in biology but presumably don't contain the same unpaid debt. Why does behavior have some miasma-like essence that confers an inexplicable quality to our being, but the structure of our kidneys does not? And second, even if humans do have some kind of special sauce, when in our evolutionary history did we acquire it? Did Neanderthals have it? Did *Homo erectus*? What about our last common ancestor with the great apes, some six to eight million years ago? Do no other species behave in a way that is inexplicable by biology? What about letting chimps into the club? Or dolphins, or the clever crows that make their own tools?

Language, perhaps the most frequently cited major difference between humans and other species, is a case in point. I will delve more deeply into the question of how communication evolved in a later chapter, but for now it is worth noting that a 2020 study by scientists from the United Kingdom found a precursor to the pathway in the brain that is required for language in monkeys, with whom humans shared a common ancestor over twenty-five million years ago.[29] Previously, scientists had imagined that the structure arose perhaps five million years ago, in a primate ancestor of humans and apes such as chimpanzees. Of course, neither monkeys nor chimps can speak. But where, and perhaps more important, when, should we declare that human language arose? What spark was emitted in time and space that signaled the transition from a brain that processed quasi-symbolic information (vervet monkeys have a special call for predators that come from above, like eagles, and another one for those that lurk on the ground, like snakes) to one that could write sonnets? The answer is that no such single moment occurred.

In the end, I think we're better off living with the sometimes messy boundaries—between species, between behavior and other characteristics, and certainly between genes and the environment or culture. In the next chapters, I will explore how behavior, with all its blurry lines and uncertain edges, evolves.

2

Snakes, Spiders, Bees, and Princesses:

How Behavior Evolves

B irds eat spiders. Snakes eat birds. Usually these two processes are unrelated, except in the deserts of Iran, rocky and barren, where a snake called *Pseudocerastes urarachnoides* lives. It looks much like any other desert-dwelling snake, with mottled scales that allow it to blend into its brown and gray background. By the early 2000s, scientists had seen a preserved specimen. They noted that it had an oddly lumpy tail with fingerlike extensions at the tip, but they didn't know whether the protrusion was a malformation in the individual that had happened to be captured, or a natural part of the animal.[1] Then they found a half-digested bird in one of the snakes. And they started to wonder.

It turned out that the knobby bit on the tail is a lure, used to attract prey. Several other snakes and lizards have body parts that resemble tasty bits of food—a worm, an insect, or other item that would appeal to something the reptile would eat. The Iranian snake is called the spider-tailed viper, because its tail not only looks but also behaves exactly, and I mean exactly, as if it were a spider, with appendages that alternately pause and scuttle like the real thing. The first time I saw a video of the snake using the lure, I was sure the filmmakers were illustrating how the tail lure worked by some-how placing a real spider on the body of a snake, an act that in retrospect seems improbable at best. But the mimicry is that good: the end of the tail has scales that are shaped to resemble the legs of a spider, each of which skitters over the body of the snake looking for all the world like they are connected to a spider body. It's not just that it looks like a spider; the tail acts like a spider, and birds apparently perceive it as one too.

In his splendidly titled blog *Life is Short but Snakes are Long*[2] (the name comes from a book review by David Quammen), Andrew Durso from Florida Gulf Coast University suggests that the tail "probably represents the most elaborate morphological caudal ornamentation known in any snake" (caudal means tail-end), and I agree, only without the qualifier "probably." But the question is, how did such a precise and elaborate mimicry evolve? Not to put too fine a point on it, but snakes just aren't that bright, and they don't have very good eyesight, and they probably don't spend any time looking at their own tails, and even if they did, they couldn't simply will a spidery appendage into existence. Furthermore, the snake cannot possibly be aware of how a spider moves, and can't know that a spider will attract prey, and can't modify its tail to behave in an ever-more alluring manner based on the response it gets. And where would the snake have gotten the idea in the first place?

Before answering that question, consider another example, one that doesn't combine quite so many phobias: bumblebees gathering pollen from flowers. Pollen is a rich source of protein for the adult bees and their young, but of course it is only available when plants are flowering. That means that if bees emerge from winter hibernation and are establishing their colonies too early in the season, before plants bloom, they risk starvation. A group of scientists from Switzerland noticed that some bumble bees were making tiny holes in the leaves of some of the plants, and that the damaged plants flowered much earlier.[3] Experiments both in the laboratory and in the field showed that the bees were much more likely to make the holes when they were starving; well-fed bees left the plants alone to flower at their leisure. This meant that hungry bees got their pollen sooner. What is more, the plants did not respond to just any holes in their leaves—if the scientists attempted to mimic the bees' damage, flowering was not accelerated nearly as much. Whether the bees have something in their saliva that induces changes in the plant's reproduction remains unknown.

A commentary on the study gave a pithy summary: "Pollen-starved bumble bees may manipulate plants to fast-forward flowering," and called the behavior "a low-cost, but highly efficient, trick."[4] A trick it may be, but no one, least of all the scientists who discovered the behavior, would suggest that the bees perform their bit of horticulture consciously: bees are hardly examining the landscape, fretting over the dearth of pollen, and selecting plants to chew on with the expectation that this will eventually yield food. So, as with the spider-tailed viper, how did such a complex behavior evolve? One could argue that the bees are exhibiting an even more sophisticated behavior than the snakes, because in the bees' case, the reward is delayed, whereas with the spider-tailed viper, food in the form of a hungry bird arrives right after the lure is deployed.

Both examples certainly illustrate the futility of expecting a "gene for" any behavior given the number of nerve cells, muscles, and other tissues and organs that are involved. There are no genes that direct agricultural activities or manifest spiders from snake scales. Even with that caveat, the evolution of these behaviors is hard to imagine. But if we think about behavior the way we would think about a physical characteristic, it is easier to approach.

First, let's review the way that any nonbehavioral characteristic evolves. Most of us understand the basics of evolution via natural selection. Imagine a population of living things—let's say a rain forest–dwelling group of birds—that is naturally variable in plumage color, with some that are greenish, some that are blue, and some that are glossy black. The variation is there initially because genes produce lots of differences, both because of how they are combined when two parents reproduce and because mutations continually arise by chance. In our case, let's say the green birds that match the trees are the most likely to evade detection by predators and survive, which means that the genes associated with green feathers are more likely to be passed on by green-feathered parents.

Eventually the population contains more green birds than the other colors, since natural selection has winnowed out the more conspicuous prey. Even after most of the population is green, any modifications that make the birds better camouflaged—say a break in the pattern so that the feathers more closely resemble leaves, or a spot that looks like a bit of decay—will still mean that the parents with better protection are more likely to have offspring. Note that mere survival via the more cryptic plumage isn't enough; the leaf-resembling birds also have to successfully reproduce, or else it won't make any difference to the gene pool in subsequent generations. Selection doesn't need to know the genes involved, just their product, as in this case the color of the plumage. And it seems self-evident as well that a bird doesn't need to know what color it is, or actively try to change its feathers.

When we talk about behavior evolving, however, the process can seem a bit more indirect. Behavior comes and goes, unlike a tail feather, so it's hard to see how individuals who did something an hour, a day, or a year ago are selected to have more babies later in life. Furthermore, behaviors seem to require agency, an internal urge to do something, and it is hard to see where that comes from. Along those lines, the Minneapolis *Star Tribune* has a regular birding column, and one week it featured nest building. In it, columnist Val Cunningham mused:[5]

> Each species [of bird] has its own nest style, and here's an amazing thing: No bird has ever observed its parents building their nest, yet in her very first season a female bird (it's usually the female) builds exactly the nest characteristic to her species. How can that be?

How indeed? The article says that it's "instinctual, hard-wired into their brains." But that doesn't answer the question about behavior, any more than saying that snakes wiggle their tails and bees punch holes in leaves because they have tail-wiggling or leaf-punching instincts. Where did the instinct come from in the first place?

A more concrete answer is that animals performing intermediate steps in complex behaviors had an advantage. In the snakes, say that the ancestors of the spider-tailed vipers had caudal lures, and already vibrated their tails when hunting, as many snakes do. The vibration seems to distract potential prey, which then attend to the tail and ignore the lethal end of the snake, to their doom. Then imagine that a few of the ancestral vipers happened to have small projections on those caudal lures, like warts. If the warty tails were more likely to attract birds than non-warty ones, the wart-bearers ate more, were more likely to survive, and had more warty babies (sorry for the mental image, particularly for the snake-averse). The snakes that happened to combine ever-wartier tails with movement got even more birds; the ones with movement that looked more spidery did even better, and so on. It's not hard to figure that some snakes are just wigglier than others; after all, even humans vary in how much they fidget, a characteristic that some scientists think is linked to our metabolic rates.

In the story about the spider-tailed viper, scientists call the prey (birds) agents of selection. The birds find certain tails more alluring than others, so they choose them, without the snakes having to do a thing. It takes many, many generations for the snakes with the spider-tipped tails to predominate, but then snakes are of a very old lineage, having arisen over one hundred million years ago. That gives them a long time for extremely small changes to accumulate.

We can construct similar scenarios for the bees that tear leaves, and for the birds that build the best nests, with similarly small variations that confer an advantage to the bearer. Bees already bite at plants sometimes, and some of the dinosaur ancestors of birds appear to have gathered materials in their surroundings when they laid their eggs. Those rain forest birds similarly grow greener and greener as their feathers accumulate the right kind of pigment. The important point is that it doesn't matter why the variant is produced—it's all about the consequences. So if a twitchy tail means the snakes eat more and are around to have more babies, a twitchy tail will appear more frequently in the population, whether the snakes are aware it's twitchy or not. The same process applies to behavior that has worked to produce extraordinarily complicated structures such as the eye (a favorite target for creationists, who often assert that such organs could not have arisen via evolution); again, a series of intermediate steps, each of which yields an advantage to the bearer, is all that's required.

Behavioral Family Trees

That series-of-intermediate-steps answer, of course, pushes the rise of these incremental behaviors back in evolutionary time without really answering the original question. Where did caudal lures come from? Why did dinosaurs build nests? And did they all do so, or just some? It is a truism that behavior doesn't fossilize and become memorialized in the geologic record, but it is still possible to understand the evolution of behavior over deep time.

First, although behavior itself doesn't turn to stone, it is still possible to infer what animals were doing in the past from their bodies as well as the ways in which their bodies are preserved. For example, we think that some dinosaurs took care of their young because groups of a single adult with several juveniles were found fossilized together. While virtually all modern lizards lay their eggs and then abandon them, the descendants of dinosaurs— birds—are champions of parenting. And fossilized animal footprints or burrows can tell a great deal about how an animal moved and what it ate, as can the structure of body parts like teeth and limbs. One group of scientists claimed that the grooves on a tooth of one *Tyrannosaurus rex* were made by another individual of the same species, which they concluded meant that the famed dinosaurs were cannibals. In a similarly gruesome example, the fossil-ized remains of a ten-foot-long snake from seventy million years ago was discovered encircling a crushed dinosaur egg in a nest of otherwise unbroken eggs. Michael Benton of the University of Bristol agreed with the authors of the study that the "snake was waiting and snatching juveniles as they hatched,"[6] which gives new meaning to the term cold-blooded.

It is also possible to reconstruct behavior by carefully analyzing fossil skeletons: an animal's stance, its jaw formation, and the rela-tive sizes of the bones in its legs can all reveal a great deal about what it ate and how it behaved. The development of 3D scanning and printing techniques has led to ever-more sophisticated models of ancient life. To return to *T. rex*, that ferocious epitome of dino-saurs, scientists, *Jurassic Park* aficionados, and six-year-olds have all long wondered how fast the predators could run. Estimates made over decades had ranged from over forty mph to eleven mph, the latter being about the speed of a human long-distance runner. Who is correct? A recent study[7] pointed out that at very high speeds, an animal as large as a *T. rex* would have risked toppling over, and furthermore that the muscles needed to power its hind legs would have had to comprise 86 percent of its body mass, a virtual impos-sibility that would have left no room for any other body functions. Hence, its likely maximum pace was close to the lower estimate, which still means the dinosaur might have been able to catch a flee-ing caveman—except that, luckily, none would be available for at least sixty-five million years after *T. rex* became extinct.

Another way to infer how behavior evolved doesn't use fossils at all. One of the most intuitive ways to understand the evolution of any characteristic is to think about its similarities in other living things, and how those similarities came to be. Evolutionarily speak-ing, objects—or behaviors—can resemble each other for one of two reasons. To illustrate, imagine an array of limbs from different animals: a whale flipper, a bird wing, a human arm, an octopus ten-tacle, and a starfish arm. Which of these is not like the others? They are all used for movement, but we know that the first three limbs share more than the same function. They have bones that are similar to each other because they are inherited from a common ancestor. In an X-ray, you can see a humerus, that long arm bone that extends from the shoulder, in each, though the shape and position are modified (and whales don't have shoulders, exactly). The bones started out the same, in an ancestral vertebrate millions of years ago, but became modified through natural selection by the particular circumstances in which each limb found itself, whether in water, or air, or on land. This type of similarity is called homology.

In contrast, both the starfish and octopus arms not only lack bones, but these animals have not shared a common ancestor for far longer than any of the vertebrates noted here. They evolved from more recent, and armless, ancestors. Selection favored extensions of their bodies in each case, but the resemblance between the limbs happened independently, through a process called convergence, or convergent evolution.

Similarity because of a common ancestor or because evolution produces similar structures through different pathways are both common. North American flying squirrels and small Australian marsupials called sugar gliders are adorable, large-eyed, nocturnal mammals with furry membranes between their front and hind legs that are used by the animals to glide between trees. But they resemble each other not because of a mutual gliding ancestor, but because of convergent evolution.

But back to behavior. People have noticed for a long time that species that look somewhat alike often act similarly as well. Take hummingbirds, for example. Many of these tiny New World birds use acrobatic aerial displays, sometimes called skydancing, to attract mates. A male hummer will ascend into the sky at high speed, then zip up and down or back and forth in a U-shaped or oval pattern. The wings—not the vocal apparatus—of the bird make snapping or whistling noises as part of the display. Each species has its own variety, with Anna's Hummingbirds flying in a tall, narrow oval;

and Costa's Hummingbirds using a much shallower path. The two species look somewhat alike, and their behavior is similar as well.

The earliest animal behaviorists, such as Konrad Lorenz, who helped develop the theory of imprinting, were fascinated by these similarities between behavior and appearance, and drew elaborate diagrams showing how such displays differ across species. But perhaps because people often think behavior is different from physical characteristics, given its fleeting nature, as I discussed in the last chapter, using such homology to understand the evolution of behavior has invited skepticism over the years. Some scientists thought behavior was just too variable to use in evolutionary studies, while others were concerned about the lack of fossilized behavior. Peter Klopfer, mentioned in chapter 1, found behavior "too malleable" to draw any conclusions about its evolution. And the late, famed paleontologist Stephen Jay Gould flatly said that "it might be interesting to know how cognition (whatever that is) arose and spread and changed, but we cannot know. Tough luck."[8]

Many of these objections, however, could equally be raised about understanding many other characteristics, not just behavior. Physiology and its associated tissue don't fossilize, or at least not very well, but we can draw many conclusions about similarities in digestion by comparing animals eating different things. Hearts are not preserved in stone, but we are pretty sure that the four-chambered variety in mammals arose from a common ancestor with crocodiles and alligators, who also have the same kind. And when you actually measure the variability in behavior, as scientists Alan de Queiroz and Peter Wimberger did[9] by examining both physical and behavioral characteristics in animals ranging from wasps to newts to birds, it turns out that behavior isn't any different from leg length or tooth size in how easy or hard it is to describe or how much it differs among individuals.

With that cleared out of the way, it becomes possible to use homology of behaviors in a different way. Instead of asking how behaviors that we see could have evolved, given a set of previously determined evolutionary relationships, we can see if behaviors shed light on those relationships themselves. It is kind of like the chicken and egg question, but with actual chickens. Or at least actual birds.

Let me explain. In this example, we'll be looking at manakins, small songbirds that live in the tropical forests of Central and South America. About forty different species of manakins exist, and the males engage in elaborate courtship displays, doing fancy dances showing off their brightly colored feathers and making sounds that, a bit like the hummingbirds, are produced by the males' wings, not their vocal system. Some species also have unusually thick bones and strong musculature associated with moving the wings. Kim Bostwick, an ornithologist at Cornell University, has studied manakins for many years. She was particularly intrigued by the Club-winged Manakin; it has an extraordinary display in which, among other acrobatics, the male turns away from the female, bends over, and shuffles backward while keeping his rear end in the air. If that sounds bizarre, well, it is, but it is also a display that has some similarities with another kind of manakin. Bostwick figured that the ancestor of the two types of manakins must have had the roots of the display, and both species then inherited versions of it.

However, when scientists looked at the anatomy of the two types of manakins, they were not that similar, and the species were judged to be evolutionarily rather far apart. So Bostwick painstakingly amassed information on both the mating displays and the bones, feathers, and other physical features of not just the two manakins in question, but of as many members of the group for which she could find good specimens.[10] Then she constructed evolutionary trees that showed how one aspect of the display, or one thickening of a bone, could be ancestral to the others, with further modifications as the species evolved. The trees revealed that the behavior and the appearance of the manakins evolved together, so that as the mating display got more sophisticated, the manakin's bones became heavier, which enabled the exaggeration of the wing noises that were used by ancestral manakins. The various parts of the display are homologous in the different manakin species, each related to the other, meaning that the behavior can be used along with the physical attributes of the bird to reconstruct the evolutionary history of the group. This combined approach is better than a history that uses only the anatomy of the birds.

Just like limbs, behaviors can exhibit either homology or convergence. Among snakes, defensive behaviors such as hoods or openmouth displays seem to be the result of convergence. In crocodiles, parental behavior is common, and is thought to be related to the care of young seen in modern-day birds, so that is a homology. What all of this means is that we can trace the evolution of behaviors in much the same way that we trace the evolution of jaws or feathers. We can also think about how quickly or slowly behavior evolves by examining how persistent behaviors are over long stretches of evolutionary time. Some behaviors, like tool use, seem to have arisen relatively quickly. Others appear to have remained unchanged for much of evolutionary history. Virtually all four-limbed vertebrates, from birds to mice to lizards, scratch their heads by lifting a rear foot over the front leg, or over the wing (humans are an obvious exception, but then we have those handy fingers). Eminent animal behaviorist John Alcock speculated that grooming behavior isn't subject to selection as animals compete and enter new habitats or stay ahead of predators—what works to scratch an itch on a mountain will do equally well in a forest and won't have an impact on the animal's survival or mating.[11] On the other hand, a behavior that allows an animal to evade attack in the desert will fail miserably in the marshes, so one might imagine that such behaviors change as selection on them changes as well.

In summary, it isn't a matter of which came first, the behavior or the physical appearance—like the chicken (or the manakin) and the egg—because you always need one to produce the other.

Scales, and a Head Full of Lizard

Evolution gives us convergence, so that structures that appear similar can have different evolutionary ancestry. It also gives us homology, so that structures that appear different can have a common origin. We also know that simpler forms gave rise to more complex ones. That's true for appearance—an amoeba is less complicated than a kangaroo—and for behavior—those spidery-tailed snake displays are more elaborate than simply waving the end of a tail back and forth.

From those principles, it's easy to develop a very common misconception, one that reveals itself in all those cartoons that show a fish sticking its head above water on the shore, followed by a reptile, then a four-footed mammal, then an ape, then a caveman (almost always with a spear), and finally a human doing something like eating a cheeseburger or typing on a keyboard. The idea of the drawing is that we are progressing ever onward, with each form more advanced than the last. Humans, then, are the pinnacle of evolution.

This belief in a hierarchical classification of living things, termed the scala naturae, is an old notion. Aristotle arranged all living beings along a scale with (predictably) humans at the top, and the other creatures beneath us in decreasing complexity. In one rendition, angels were included, and were seen as closest to God, followed by humans, and then other animals. Similar ideas have been perpetuated over the centuries, with one of the more recent versions being a "ladder" of evolution, so that again, humans are at the apex, preceded by other mammals, which are themselves preceded by reptiles, fish, and on down to the invertebrates, each with its own rung. It is as though living things are in a gigantic military, with microbes or worms as the privates and people as the generals.

Persistent though it is (one can still find references to an evolutionary ladder in some modern textbooks), the scala naturae is not just outdated, it is completely erroneous. It is true that some groups of organisms, including humans, arose more recently in evolutionary history than others. But recent evolution isn't an award, it's just an attribute. Domestic dogs arose more recently than wolves, but also more recently than humans. The novel coronavirus arose more recently than leprosy, and both of them are newer than people. So who is at the top? Evolution leads to a bush, not a ladder, with the branches of the bush representing change from one form to the next over time.[12] The tips of the branches can be thought of as the living things that are present now, each one as highly evolved as the other.

The scala naturae rears its head when it comes to behavior, because it can make us think that if everything is always getting more complex, then behavior must exist on a continuum, with simple actions like the tumbling bacteria giving way to a tail-wagging dog, a nest-building bird, and eventually, a brain-surgery-practicing human. And since the brain ultimately produces our behavior, it can be tempting to see behavior evolving the same way, as if mammal brains are like reptile brains, but with added complexity. If so, then as newer, more sophisticated parts of the brain evolved, they were layered onto the older ones. It's like the notion of human ice cream bars I alluded to in the last chapter.

This concept might sound familiar to anyone who has heard—or made—a reference to a "lizard brain" that makes decisions based on emotion or instinct rather than reason. Called the triune brain theory and developed by psychologist Paul MacLean in the 1940s and 1950s,[13] the lizard brain was also popularized in Carl Sagan's famous book *The Dragons of Eden*.[14] MacLean postulated that the modern vertebrate brain had three units: an atavistic reptilian one that only takes itself into account, an early mammalian one that is emotional but potentially unselfish, and a more rational later mammalian one. The interaction of these components produces the often-contradictory behaviors and impulses we see enacted.

As neurobiologist Anton Reiner noted in a review of one of MacLean's books,[15] the triune brain idea has a lot of appeal. It allows us to classify less-desirable behaviors, like flying into a rage at the checkout line in the supermarket, as somehow nonhuman, and perhaps therefore something we are not responsible for. It helps fuel that sense of human exceptionalism I discussed in chapter 1, as though even our brains represent the latest, most effective model, the one that superseded the clunky old-fashioned brain our distant ancestors were forced to use. It encourages us to dismiss "instincts" as a holdover from our past, separate from a more reasoned way of thinking. And, as Reiner says, MacLean's ideas "are also appealing because they are simple; after a ten-minute exposition of them one can feel equipped to explain much human behavior with the force of science behind one."

The problem is that the triune brain, like the scala naturae, doesn't exist, or, as the title of a 2020 paper puts it, "Your Brain Is Not an Onion With a Tiny Reptile Inside."[16] (Whether you find the onion or ice cream bar metaphor more appealing is, I imagine, a matter of personal taste.) Modern neurobiology shows quite clearly that brains do not possess "newer" parts affixed onto "older" ones, and furthermore, that brain and behavioral complexity do not map onto any kind of evolutionary sequence. And we cannot assign whole classes of behavior, such as territoriality or aggression, to a particular part of the brain, since the brain and nervous system are much more integrated than the triune brain model implies. Finally, MacLean considered parenting to be part of the more advanced behavioral repertoire, and hence associated with the mammalian part of the brain, even though crocodiles, birds, and at least some dinosaurs took care of their young.

Yet the model has been hard to debunk, even among psychologists. The authors of the brain-is-not-an-onion paper, led by Joseph Cesario at Michigan State University, surveyed recent introductory psychology texts, and found some version of the triune brain misconception in nearly 90 percent of them.[17] Cesario and his coauthors point out that contrasting an ancestral, impulsive, "animalistic" nature with a more long-term rational one lies behind research on willpower that touts the ability to forego reward with a more mature outlook on life. Instead, they argue that sometimes choosing the immediate reward is more beneficial than waiting; it all depends on the circumstances. Thus, they say, "The question of willpower is not 'Why do people act sometimes like hedonic animals and sometimes like rational humans?' but instead, 'What are the general principles by which animals make decisions about opportunity costs?'"

Getting to the Genes, but Not the Way You Think

All of this talk about fossils and brains and similarities among different kinds of animals leaves out the link between genes and behavior itself, a link that obviously exists for behavior just as it does for physical characteristics. I will explain how we know which and how many genes are associated with specific behaviors in more detail in the next chapter, but before I do, it is worth noting just how indirect the connection can be between any one, or even many, genes and the behavior that we observe.

Let's start by thinking about dogs. People quite happily acknowledge that while training and obedience classes can shape a puppy's behavior, some of that behavior also comes from its breed. Retrievers are called that not because of how they look, but because of what they do. Dog owners are happy to wax rhapsodic about the behavioral quirks of particular breeds. The American Kennel Club[18] acknowledges "personality" as one of the hallmarks of recognized breeds, and includes "may be stubborn" and "eager to please" as one of the search terms one can use for finding a pet, alongside "infrequent shedding" and "small size." In fact, the behavioral selection criteria—for activity level, propensity toward barking, and trainability—outnumber the ones for physical characteristics.

Of course, people in the past who selected for specific traits in dogs did not know anything about genes. They just kept choosing the puppies who shed less or barked more, and kept doing so until those traits became pronounced in the breed.

At the same time, it is illuminating to understand just how convoluted the path from gene to behavior can be. As an illustration, consider a breed of dog called the Australian kelpie, a "working dog" that herds sheep and cattle. Kelpies are capable of doing their job with relatively little guidance, which is useful in the vast Australian outback. Like other breeds of domestic dogs, humans selected the individuals that behaved in a certain way—kelpies need to unhesitatingly muster the group of animals they are looking after, and they need to keep at it for hours at a time, often without food or water.

Claire Wade, a professor in animal genetics at the University of Sydney in Australia, has been studying genes in kelpies for many years. Using new technology, she can compare the DNA of kelpies that work as herders with those that are kept as pets, and she can also compare both to other breeds of dogs. It turns out that a crucial difference in the genes of the working kelpies is a section of DNA that is associated with, of all things, pain tolerance. The working dogs have higher pain tolerance than the other dogs. How could that lead to better herding?

Wade points out that "the ability to feel pain is a stop|go requirement for working Kelpies. In outback Australia, the ground coverings are extremely prickly. I always tell people of a story where I was visiting a friend and my dog ran out into the field but then froze and would no longer move. I needed to go and carry her back to the soft grass—we call that being a 'prickle princess.' Kelpies cannot be a 'prickle princess' or they never have a chance to demonstrate their ability for moving sheep."[19]

So the dogs that felt less pain—because of a variant in their nervous system and the way that messages are transmitted among nerve cells—were the ones that were chosen for the breed. There is no gene for herding per se, but herding evolves nonetheless. Dogs that were not "prickle princesses" had more puppies than the ones that were carried out of the field, and so they became kelpie ances-tors. Presumably, this kind of indirect manner of selection works for many creatures, not just domestic animals.

Does Behavior Lead and Evolution Follow?

Behavior may be continuous with morphology, and it may evolve like morphology, but we still recognize its fleeting nature. A dog wags its tail, a cheetah sprints after a gazelle, and then it's over, the friendliness conveyed, the prey obtained or missed. Unlike a body part, behavior vanishes. How does the wagging or the sprinting get incorporated into the genes, a prerequisite for its evolution? Fur-thermore, many behaviors are at least partly learned, from the local dialect of a White-crowned Sparrow to the way a chimpanzee holds its termite-collection stick. So how does that learning fit into the evolutionary process?

Biologist James Mark Baldwin suggested that if an animal learned to do a new thing, and that new thing helps it survive and reproduce better, its genes are more likely to be passed along, even though there is no direct connection between the task that is learned and any particular gene or set of genes. This doesn't mean that the animal wills evolution into happening, but, as the eminent animal behaviorist Sir Patrick Bateson put it, "Whole organisms survive and reproduce differentially and the winners drag their genotypes with them."[20]

You could also think of this as behavior leading the way for evo-lution. When the environment changes, behavior, being inherently flexible, is how an animal first responds. That response then makes it possible for the animal to become adapted to its environment. This idea has been somewhat controversial, perhaps because as has become clear by now, behavior, physiology, and physical attributes are all very tightly linked. It is therefore difficult to pinpoint one of them as the obvious starting point.

These exchanges between behavior and the environment can be seen most vividly in the case of something rather stuffily termed niche construction. Ecologists refer to a niche as the sum of all the requirements of an animal (or plant) necessary for it to survive. A monarch butterfly, for example, needs to have milkweed to eat as a caterpillar, and the temperature must be above 55 degrees Fahrenheit for it to be able to fly. A mole has to have soil of the right consistency to burrow into and a nice selection of insects to eat, as well as a range of temperatures that are neither too hot nor too cold. Add up all those necessities, and they define the niche.

But what if the niche itself is influenced by the animal's activities? One of the most well-known examples of such influence is that of beavers making a dam. A new dam starts when beavers bring tree branches to a stream and set them in such a way that the flow of the water is decreased. After the base is constructed, plant material and mud or rocks are used to build up the structure, with the beavers moving entire logs through the water. Eventually, the dam causes the section of river to flood, making a pond, and the beavers then make the lodge there to live in. Beaver dams are impressive constructions: beavers can move trees up to three feet in diameter to build them, and their dams can occupy several hundred feet of the habitat. The resulting pond provides protection from predators such as coyotes, and the dam also serves as a source of food in the winter, since beavers eat bark and woody plant materials.

Once constructed, the dam alters many things about the stream or river in which it is built, including its flow patterns, the way that leaves are decomposed in the water, and the kind of plants that can thrive on its edge. This means that the food sources of the beaver change, along with the habitat for birds, fish, and insects. Eventually, the whole area looks different because of the beaver dam. The beavers' offspring then inherit the dam and its surroundings, which in turn means that natural selection acts differently on the generations that grow up with the altered environment than it did on the originators of the dam. That changed environment in turn has the potential to influence the evolution of the beavers' behavior, and so forth.

Darwin and Emotional Evolution

Finally, what about the evolution of perhaps the most mysterious of behaviors, emotions; or as animal behaviorist Gordon Burghardt puts it, the "private experiences" that we, and perhaps some other animals, have?[21] I will explore the evolution of intelligence and cognition in later chapters, but here it is worth thinking about just how something as hard to define but as important (at least to us) as an emotion could evolve.

Charles Darwin was extremely interested in how behavior, and emotions, could evolve. He was particularly fascinated by the idea that we could trace similarities in behavior across different kinds of animals much the same way we could see resemblances in their bones or teeth. His *The Expression of the Emotions in Man and Animals*[22] is about the ways that animals show emotions, such as fear or anger, reflecting our common heritage, and has the now-famous illustrations of similar facial expressions in humans from across the world as well as in apes. He performed an experiment, advanced for its time, in which he showed guests at his home a series of photographs of faces representing different emotional states and asked them to identify the emotion depicted. (You have to wonder how this affected the likelihood of people accepting his dinner invitations, not to mention what his wife Emma thought of the plan.) Although the guests agreed on some of the emotions depicted in the photos, including happiness, sadness, fear, and surprise, they strongly disagreed about the more ambiguous emotions such as jealousy that the images showed. Darwin took this to mean that only certain basic emotional states are universal, as he had theorized. He further interpreted animal communication as a means for conveying emotions, so that signals like a dog raising the hair on its neck could have come about as way to show fear because it was connected to what he called "the direct action of the nervous system."

Biologists viewed Darwin's work on animal emotions with unease for quite some time. Even well after his ideas about evolution by natural selection were accepted, many found his speculations on emotion to be anthropomorphic at best and downright squishy and embarrassing at worst. As Burghardt noted, "The reaction from human chauvinists in biology, the social sciences, and the humanities was swift and often brutal."[23] Klopfer went so far as to call Darwin's enthusiasm for research into emotions a "scientific dead-end" and titled his review of a book attempting to reexamine research into animal emotions "Still Largely Where Darwin Left Us."[24]

Part of the difficulty was and is that although people are fine— perhaps too fine, as I will argue in a later chapter—with seeing their dogs as being capable of jealousy, rage, or grief that is identical to that of humans, other animals are a harder sell. Elephants may mourn their dead, but what about, say, snakes? Tropical biologist Alexander Skutch said that snakes are creatures "in which we detect no joy and no emotion."[25] This is a sentiment with which many people would agree. But why? Do you have to have facial expressions to have emotions? Or at least eyelids, or lips?

The eminent primatologist Frans de Waal is a passionate proponent of the idea that nonhuman animals have emotions that are virtually identical to those of humans, although most of his work has been with chimpanzees and bonobos, creatures in which it is easy to see human behavior reflected. His book *Mama's Last Hug*[26] is filled with examples of animals exhibiting what certainly looks like jealousy, shame, and compassion. He also distinguishes between feelings and emotions. The former, he says, are the internal states that only the individual experiencing them can truly know, while the latter are "bodily and mental states . . . that drive behavior." The emotions, De Waal suggests, are easily noted in other species. He finds the lack of willingness to acknowledge the similarities between humans and other animals a kind of arrogance, a form of the human exceptionalism that I noted in the last chapter.

I am sympathetic to the objection to human exceptionalism, and I certainly agree with De Waal about the continuity and the relationship between humans and other animals. At the same time, the solution shouldn't be to lump us all together. It seems to me equally anthropocentric, perhaps even narcissistic, to assume that all—or any—other species experience life or emotions exactly the way that humans do. I study insects, and nothing is as humbling as the realization that I have very little idea about their emotions, or whether they have any at all that I could fathom. Their brains and nervous systems are completely different from our own—no homology there. And if I treat them like little people in exoskeletons, all I learn is how well I project my own emotions onto other beings.

It is tempting to start making categories here: primates are like people, and maybe along with all mammals, they should feel some form of emotions, so they go in the basket with humans. And maybe we should add crows and their kin, whose abilities I extoll in a later chapter. I realize that most people are fine with leaving me to neglect the emotional lives of insects. But eventually this sorting of creatures becomes unwieldy, and it also brings us back to that scala naturae, with a ranking depending on closeness to humans that then means some animals get to have emotions and others do not. But the choices are not all or none, human replica versus robot that feels nothing.

And with regard to those robotic insects, a 2020 article in the *New York Times*[27] reported on a study of praying mantises, highlighting the insects' ability to adjust their lethal strike depending on the speed of their prey. In a slow-motion video of a hunt, according to the author of the article, "We see the mantis pause and calibrate, almost like an experienced baseball catcher who has realized she's dealing with a knuckleball."

The subtitle of the piece is "New Research Shows These Ferocious Insects Don't Just Hunt Like Robots." Leaving aside exactly how we know the way robots hunt, the reaction of the readers to the article reveals a great deal about how we do, or don't, see emotions in other species:

> Why we should think that they, or any other living entity oper-ates as if a robot is way beyond me. Do we really need to bring them into a lab to realise that they too are conscious, feeling beings?

> Sorry to break this to you, fellow humans, but every pig you eat, and every bug you crush, had a life, a will to live, a sense of self, the capacity to feel fear, relief, hunger, lust, etc. Other animals may not have the same emotional range we have (although, who knows), but it's likely that they experience some things, e.g., terror, even more intensely than do we . . . Praying mantises have awareness and can think, and are not, in fact, robots—any more than are ants, ant-eaters, hippos, worms, bees, snakes, or donkeys. So treat them accordingly, whenever possible.

If it is self-centered to assume we are the only animals to think and feel, why is it not equally self-centered to assume they think and feel as we do? The funny thing about seeing all animals—or all mammals, or all vertebrates—as sharing the same emotions as humans is that we have no such expectations of their many other characteristics. Hummingbirds' hearts beat twenty-seven times faster than humans. Whales can dive to depths of two thousand meters on a single breath. Cats do not ovulate until after they mate. It is easy to come up with examples of animal functions that are vastly different from those of our own. And if it is reasonable for other animals to have vastly different reproductive or respira-tory systems than people, why should we think they have the same emotions that we do? We should grant behavior the same evolu-tionary courtesy we do other characteristics, and try to under-stand how it sometimes came to be different in other species and sometimes the same.

3

Clean-Minded Bees and Courtship Genes:
The Inheritance of Behavior

E volution depends on characteristics being passed from parents to offspring. Behavior, like other traits, is the result of a combination of influence from the genes and from the environment. But while it is relatively easy to understand how the environment changes behavior, the connection between genes and behavior is a little harder to comprehend.

Of course, one doesn't have to know anything about genes to think about evolution. Darwin was able to develop his theory of how characteristics—including behavior—evolved without knowing the mechanism of how parents pass traits to their offspring. It took the monk Gregor Mendel, working in his garden in the Augustinian St Thomas's Abbey in Brno, now part of the Czech Republic, to find the key to inheritance. During the mid-nineteenth century, Mendel used careful experiments to show how pea plants could pass on smooth or wrinkled seeds to their offspring. Working out the proportions of each type in the seedlings of a particular set of parents, he was able to show that each parent plant had an element that it contributed to the seed, and that those elements stayed as distinct entities in the seedlings. When one crossed a smooth-seeded parent and a wrinkly-seeded parent, the seedlings were either smooth or wrinkled, but not somewhere in between. This notion contrasts with the common view at the time, which was that the characteristics of the parents were always blended together, as if one were mixing paint, so that a red parent and a white parent (to switch metaphors) would have pink offspring. We now know that traits that follow Mendelian inheritance are due to one or a few genes together.

Determining the nature of genes was a huge breakthrough, and once Mendel's work was recognized in the early twentieth century, scientists began to study which traits were associated with single genes. The early history of genetics is filled with discoveries about the genetic basis of everything from eye color in fruit flies to the number of kernels in an ear of corn. Not all the traits were inherited in a Mendelian fashion, as I discuss later, but many were.

A few behavioral traits are inherited almost as simply as seed texture in Mendel's peas. For many decades, beekeepers have been vigilant about a nasty disease called American foulbrood, which invades beehives and eventually kills the larvae inside, decimating the colony. The queen bee lays eggs that hatch into larvae inside the hexagonal wax cells of the hive. The workers will then rear the young bees. A diseased larva can quickly spread its infection through contact with the bees in the colony tending to it. But some colonies show what is called hygienic behavior: adult bees tending the young will recognize a sick larva, saw through the top of the wax cover of the cell, remove the diseased individual, and take it away from the hive, preventing further transmission.

In the 1960s, scientist Walter Rothenbuhler did experiments on the control of foulbrood similar to those performed by Mendel on the peas.[1] He mated bees from hygienic colonies with bees from colonies that were not hygienic, and then examined the behavior of the colonies that were produced. According to his results, the hygienic behavior depended on two genes: one that controlled the bee's propensity to uncap the wax, and another that caused the bees to remove the larva, limiting the spread of the infection. Thus, in the "hybrid" bees (I put hybrid in quotation marks because the bees were all the same species, just with different behaviors), some were completely hygienic—meaning that they successfully rid their colonies of foulbrood—and some were not. The non-hygienic bee colonies were of two types. In some, the bees could remove a larva if the beekeeper uncapped the cell first, while in others, workers would uncap the cell, but then leave the larva rotting away. Each behavior was associated with one gene, and both genes were needed to complete the entire sequence. Since that time, more genes have been discovered that moderate the behavior, and colonies have been found to vary in the extent to which they enact the removal of the diseased young, but it is still a relatively simple correspondence between genes and action.

Rothenbuhler's work couldn't identify the genes that were responsible for the behavior, largely because well into the twentieth century, scientists knew little about genes in most animals. The necessary delicate microscope work and tedious experiments made only a few living things possible subjects. During the early 1900s, most genetic research relied on a single animal, *Drosophila*, commonly called fruit flies. No one would call them charismatic—they are the tiny brown creatures attracted to rotting bananas in your kitchen—but they were of immense importance in the development of genetics in the early to mid-twentieth century, and they continue to be essential animal models for studies ranging from genetics to physiology. *Drosophila* are easy to rear in the laboratory, they have chromosomes that can be seen more easily under the microscope than those of other organisms, and they can be bred to show different genetic changes, or mutations, in their physical appearance. In the days before DNA sequencing, or even before DNA was known to be the stuff of genes, much progress was made by breeding fruit flies with obviously different traits and analyzing the resulting offspring.

However, although scientists diligently placed fruit flies with different eye colors or wing shapes together in glass bottles (the preferred way to harbor the insects) and counted the types of juveniles produced from these crosses (as they are called), for a variety of reasons, they rarely if ever bothered to watch the fruit flies mating. As an aficionado of insect sex myself, I find that mind-boggling. How can you not wonder about the drama of courtship and mating evaluation that constitutes sexual behavior? Reluctantly, however, I admit that not everyone, including other scientists, shares my enthusiasm.

A landmark exception was Margaret Bastock, a PhD student working under the illustrious ethologist Niko Tinbergen at the University of Oxford in the 1950s. Tinbergen and most of his group studied vertebrates, mainly fish and birds, but Bastock wanted to use *Drosophila* to see how the relatively new science of genetics could help explain differences in behavior, and in turn, the way that behavior could shape evolution.[2] She decided to use a mutant called *yellow*, caused by a single gene which, as one might imagine, renders the fruit flies golden colored. These mutant flies also occur in the wild, meaning that one can collect *yellow* fruit flies at garbage dumps and other places where the flies congregate, but *yellow* fruit flies are more common in the laboratory. Earlier researchers had seen that the male *yellow* fruit flies fathered fewer offspring, but didn't know why.

Bastock first arranged some matings between the normal "wild type" (the *Drosophila* ordinarily found in nature) and *yellow* individuals. She reared their young to create groups of fruit flies that differed only in the *yellow* gene. She then proceeded to perform a series of exquisitely detailed experiments, including observations of courtship and mating behavior in which she sat in front of a pair of fruit flies and spoke into a microphone to record, on a reel-to-reel tape recorder, what the fruit flies were doing every one-and-a-half seconds. It is worth pausing for a moment to think about this less glamorous part of watching animals behave: peering at fruit flies under a microscope for hours on end is not much like watching lions in the Serengeti, though of course that can be tedious too.

Part of an amorous male fruit fly's courtship includes vibrating his wings and licking the female while he positions himself behind her. The female has to cooperate for mating to occur, and males will often persist for long periods before either giving up or getting a chance to mate. Bastock found that *yellow* males did all of these courtship behaviors less often and for less time than the non-*yellow* flies, and that this lowered vigor seemed to be what was responsible for the difference in success at reproduction. Bastock wondered if the reason the males were lackluster was that the females did not react to them the same way they did to the wild-type fruit flies, rather than because of something in the males themselves. She did yet more observations to find out. Nope, the females employed the same behavior for all of the male flies. It really was, as she said in the title of her paper, "A Gene Mutation Which Changes a Behavior Pattern."[3]

This was a big deal. No one had demonstrated a clear link between a known gene and what an animal did. What is more, Bastock pointed out that such genetically driven behavioral differences can play an important role in that most important of evolutionary processes, the formation of new species. As she recognized, if some of the males in a population of flies stopped vibrating their wings as much, perhaps some females would turn their attention to other features of courtship, like odor, or the pattern of spots on the wings. That in turn would mean that males with the newer method of attracting a mate were favored by selection, which could lead to the separation of these fruit flies from those that followed the more old-fashioned technique. Eventually, the populations might diverge so much that females from one group would not mate with males from the other, and vice versa. All because of a single gene that affected behavior.

Plastic Expressions

Since Bastock's time, of course, research into the role of genes influencing behavior has exploded. But the novelty of her finding endures—it is extremely rare to find just one gene that has a big effect on complex behavior. And when it happens, it's often not the presence or absence of a gene that is important, as it was with the *yellow* fruit flies, but a change in whether or not a gene is expressed.

To understand the distinction, remember that a gene is just the term we use for the DNA chunk that sits on a chromosome (there are actually philosophical debates about just what a gene is, but we'll set those aside for our purposes). How does the DNA end up making a muscle fiber, or a nerve cell, or any of the other parts of the body? The answer is that the genes have to be expressed, which means that first the information they contain must be made into a different genetic material and then into a final product, like the aforementioned muscle fiber. Crucially, different genes are expressed at different times, which is why you have had the same genes all your life but only started growing body hair at puberty. It is a complicated process, and one that modern techniques have only recently allowed us to explore.

Back to behavior, and this time to a different set of insects: ants. Like honeybees, ants are highly social. Most ant species live in colonies with a queen that does all the reproducing and workers, all female, that do pretty much everything else—find food, care for the larvae, and defend the nest. The queen and workers are often genetically identical. So what makes a queen, a queen?

To answer this question, Daniel Kronauer from Rockefeller University in New York and his colleagues compared gene expression in seven different species of ants, some with queens and some that live in groups of workers that reproduce without mating, making what are essentially clones of themselves.[4] The queenless groups appear to have arisen more recently in evolution, though like other ants, they cooperate to care for the larvae. Across the species, one gene was activated to manufacture more of its product in the reproducing individuals of a colony than in the workers: insulin-like peptide 2, or ILP_2. This means that the most recent common ancestor of ants likely had high ILP_2 expression in its reproducing individuals and low expression in its workers.

Then the researchers looked at the clonal raider ant *Ooceraea biroi*, which is one of the queenless species. They removed the larvae from colonies that were in their caregiving stage, and found that the expression of ILP_2 went way up within twelve hours. Conversely, if ants were offered larvae, the expression of that gene went down. What is more, when they injected the peptide into workers in colonies with larvae (a procedure that requires a steady hand, a lot of practice, and a very tiny needle), their ovaries were switched on, as if they were about to reproduce. The higher the dose, the more eggs developed inside the workers' bodies.

This insulin-like peptide is similar to the insulin found in the human body, and indeed in many kinds of animals. In the ants it appears to increase the likelihood that they will go foraging and then give the food to their larvae. An article in the *New York Times* about the research drew the analogy to people getting hungry when their insulin levels fall, although in the ants presumably that doesn't induce preparing macaroni and cheese for the larvae.

In addition to illuminating a tiny part of how a complex social system like that found in ants could have evolved, the ILP_2 story shows once again that genes by themselves don't induce behavior. Having high or low levels of ILP_2 activation means nothing without the presence or absence of larvae in the environment. At the same time, you can't just make larvae without the necessary genetic machinery that switches on the ovaries.

Knowing more about gene expression also sheds light on the question of how new behaviors get incorporated in the genome. Work by renowned bee genetics expert Gene Robinson might have some answers. The African honeybee is very similar to our more familiar variety. It is famous, however, for its aggressive defense of the hive, as people have discovered when the African bees have been inadvertently introduced into places where people and bees come into contact. Robinson and his colleagues discovered[5] that, as compared to other honeybees, the brains of the African bees show greater expression in genes that make them respond to the alarm pheromone, the chemical that signals danger to the hive and triggers the colony members to seek out and sting intruders. This means that the degree of ferocity, so to speak, is flexible, depending on gene expression, but not on the presence or absence of new genes.

Why might that flexibility have evolved? Imagine bees in an environment with many threats to the hive. In such a scenario, a colony with denizens that had greater hair-trigger responses to alarm pheromones would be more likely to survive than those with a more lackadaisical response. Thus, colonies with higher levels of gene expression would do better, which would then translate into an overall heightened level of aggression in that type of bee. Same genes, but different activation levels.

On Height and Lovebirds

Most traits, whether behavioral or not, are difficult or impossible to attribute to one or a few genes, and that is true in animals as well as people. Instead, they arise from a complex interaction (that word again) between many genes and the environment of an individual. The most common example of such a characteristic is how tall you are. Your height obviously reflects the heights of both your parents as well as the kind of environment you had as a child: if you were malnourished, you did not grow to be as tall as if you had an adequate diet, even if your parents were both above average in height. Many different genes contribute to height, which means that there are many different heights in any one population—if your mother was five feet four inches and your father five feet nine inches, your height might be the same as one of them, somewhere in between, or even shorter or taller than either of them.

This kind of distribution is common in animals as well, much more so than the single or few-gene associations illustrated by the bees, who either uncapped cells or didn't (no bees uncapped a cell part of the way and then left the rest of the cap alone, or sniffed at a diseased larva and then did not attempt to remove it). That either-

or dichotomy is rare. Instead, behaviors are more likely to occur along a spectrum, like height in humans.

Consider lovebirds, those African parrots often touted as paragons of monogamy. Different species of lovebirds can be bred with each other in captivity, but each species also has its own set of distinctive behaviors. William Dilger studied hybrids between two kinds of lovebirds in the 1950s and 1960s.[6] He clearly harbored real affection for his study animals, reflecting in an article: "The partners exhibit their mutual interest with great constancy and in a variety of beguiling activities." Both the Masked Lovebird and the Peach-faced Lovebird nest in tree holes, and both bring bark and grass back to the trees to use as nesting material, but they do so in completely different ways. The Masked Lovebird carries those materials in its beak, while the Peach-faced Lovebird carries grass by tucking it into its rump feathers. The hybrids that were the result of a male from one species mating with a female from the other acted, as Dilger put it, "completely confused." They showed a range of behaviors—they might begin to carry one strip of material in the beak and then stop before they get to the nest, or they might carry several in the rump feathers and then lose them all partway. Unlike the bees, the lovebirds did not exhibit the behavior of one or another of the parental species, but did things that had not been observed in either. Interestingly, over time the nest-building behavior improved, so that after a few years, the hybrids successfully managed to build a structure that worked as a nest. Whether that was because they observed other birds or simply learned from their own experiences is unclear.

That kind of variety of behaviors in the offspring of a hybrid means that many genes are involved in influencing the behavior. If it were only one or two genes, we would see just one or two forms of behavior, as with the hygienic bees, and nothing in between. And the effect of learning shows that, once again, behaviors, no matter how rigid they may seem, are not produced in a vacuum.

What now? If many genes influence a behavior, and the environment, or culture, or learning all affect the behavior as well, is that it? Do we simply say, "It's a lot of genes, and it's complicated" and leave it at that? We could—and sometimes I think the truism that "things are complicated" gets less credit for its profundity than it deserves—but we don't have to, because farmers got there first, and their discoveries have helped us understand how behavior can evolve even when many genes are involved.

Well, it is not entirely true that farmers made the discovery. But the interest of agriculturalists who wanted to get faster horses, bigger ears of corn, or cows that produced more milk helped motivate scientists and mathematicians to develop a way to numerically measure the degree to which ancestry mattered. As Darwin's ideas about evolution were developing into their modern version, biologists argued about how heredity might work. They had read Mendel, but also acknowledged that everything was not inherited the same way. Farmers didn't care about Darwin, but were interested in which plants or animals to use as seed or breeding stock, because it was important for them to know how much change they could expect after a certain number of generations. Breeding a cow that produced a pint more milk per day would be wonderful, but it would be of limited use if it took a thousand generations to get there. Eventually, the scientists developed a method to describe the way that differences in traits like height, which show a wide range from short to tall, could be attributed to genes and hence could likely be successfully selected to improve.

We call height a "quantitative trait," and it and others like it can be studied using an approach called quantitative genetics. Rather than looking at individuals, quantitative genetics examines populations and the variation in characteristics that they contain. Why, for example, are Icelandic people taller and Argentinians shorter relative to other groups of humans? The answer is that they are genetically different, at least in part. But more than that, quantitative genetics asks about the variation within a group of individuals. Every Icelandic person isn't the same height, and neither is every Argentinian. What is more, we know that if a given Argentinian has a poor diet as a child, he or she might end up shorter than if the same person were well fed. Using techniques originally developed for animal and plant breeders, it is possible to measure how much genes or the environment can explain variation in a trait, whether behavioral or not. The amount of variation that can be explained by the genes is referred to as "heritability," a term that is so often misunderstood that virtually every discussion of it bemoans the confusion. Heritability is measured as a percentage or proportion, so that a given characteristic, whether height or anything else, can be assigned a number as a percent, between 0 and 100, or, if you prefer, a proportion, between 0 and 1. But that score is not something you carry around with you, or can use to describe yourself. In fact, it isn't a characteristic of an individual at all.

Part of the problem is that people tend to gloss over the words *variation* and *population* and go straight for the "how much is genes" part. But that is a mistake. To illustrate, let's think about plants, as the Harvard University geneticist Richard Lewontin did in his classic example of the concept.[7] Imagine that you take a group of basil seedlings and plant them in two trays of soil. You put one tray in a place with sun, and you water and fertilize the seedlings as they grow. The other tray is left in a gloomy corner of your yard, and you hardly ever remember to water it. Both groups of plants eventually mature, and the plants in the first tray are, on average, taller than those in the second tray. What is the heritability of plant height? You can't answer that question right away, and no answer would pertain to all the basil plants. First, you have to know the variation in height among the plants within each tray. In the first tray, say that some plants are ten inches high, some are twelve, some are fourteen, and one each is eighteen and six inches. Why are the differences there? The heritability *within that tray* is close to 100 percent; the variation among the individual plants only occurs because the seeds each had different genes, since they were all in the same environment. Note that I said "close to 100 percent." That is because it is virtually impossible that each seedling truly experienced the same environment. Perhaps water was more likely to pool at one end of the tray, or perhaps the seedlings at the corners had somewhat more room. Still, it is reasonable to conclude that the variation among the plants—not the height of any single individual—can be attributed to their genes.

Now let us turn our attention to the second, neglected, tray. In contrast to the first, say that its inhabitants are between four and ten inches tall. Once again, we can measure the heritability by examining the variation among the individuals *within the tray*. And we can also determine that the difference between the trays is because of the difference in their environments. However, we can never talk about heritability as an individual characteristic, and we can never talk about it separate from the environment in which we measured it. People sometimes say that "heritability is a local measure," emphasizing that the degree to which genes account for the variation in a trait depends on the circumstances when it is measured. This also means that heritability can increase or decrease: if you reduce the variation in the environment that an organism is in, the remaining variance you see in the trait you are measuring must be due to the genes, and since heritability is a proportion, the number describing that gene variance has to go up.

The Power of Two, and Evolutionary Bookkeeping

The plant example is clear in part because the experiment is so controlled. But we can calculate heritability in animals, as well as in people, and it is perfectly possible to determine heritability in behavior as well as in physical attributes. In humans a common way of doing so is to take advantage of the genetic similarities in twins. As you probably know, human twins come in two basic flavors: identical, which means they started out as a single fertilized egg that then split, with each of the resulting halves genetically the same as the other; and fraternal, which are like ordinary siblings in that they arose from two different eggs fertilized by different sperm.

Scientists recognized the power of examining twins many years ago, and my own university, the University of Minnesota, has a famous place for doing just that, the Minnesota Center for Twin and Family Research.[8] The basic idea is simple: if we compare identical twins that were raised in different environments, perhaps because each member was adopted into a different family outside of the birth family, then we can see if they grow up to be more like each other or if they become more like the siblings in their respective adoptive family. If the twins are different, that difference may be said to be due to their different environments. Similarly, we can compare the traits of fraternal twins raised in the same environment: if they are then different, that difference is due to their genes, not the environment.

Everyone, including the Minnesota Center, recognizes that these broad generalizations are just that. The research is quite complicated, using repeated questionnaires, family history data, sophisticated statistical modeling, and more. Since its start in 1989, the Center has gathered information from over 9,800 people to date, and they have examined behaviors ranging from eating disorders to personality traits to happiness. The results have been illuminating in many ways, perhaps most by acknowledging the importance of both genes and environment in a wide range of characteristics. For example, even fingerprints are not identical in identical twins, though they are more similar in the pair than they are between fraternal twins. And attributes such as general levels of happiness or a tendency to some personality disorders, while more similar in identical twins, are also influenced by family life. Sometimes genetic influences change over time.[9] The twins' concerns about body shape and weight were not as affected by genetic similarity in preadolescent twins as they were in twins from early adolescence onward. Twins understandably fascinate us, but their similarities don't always mean what people think, as I detail later on.

Animals don't lend themselves to twin studies in quite the same way, but in many species the offspring are born or hatched in groups, and while the siblings are not genetically identical, they still share more of their genes than a random pair of individuals. To measure heritability of a characteristic in animals, one can approximate the scenario of a twin raised by a family other than the one he or she was born in by doing what is called a cross-fostering study. These are most frequently accomplished using songbirds, with their handy nests full of chicks as well as their general inability to recognize their own offspring. (Side note, and public service announcement: the idea that if you touch a baby bird on the ground and put it back in its nest, its mother will reject it because it "smells of human" is a myth, perhaps perpetuated by overworked mothers who did not want their children messing with chicks in the first place. Please, by all means, if you see the nest, put the baby back—the mother, like most birds, has a rather poor sense of smell and will feed it again without skipping a beat, or a worm. Or you can call a wildlife rehabilitation center.) For cross-fostering, a researcher selects two nests with eggs of similar age and waits for the chicks to hatch. Once they do, half of each brood is swapped between them, so that the parents of nest number one raise half their own chicks and half from nest number two, and vice versa. The researchers can then measure the characteristics of the chicks in each nest and compare them to the parents that reared them and to their genetic parents.

One can also do an animal heritability study experimentally, depending on the species in question. My one experience with such a study used Red Junglefowl, the ancestor of domestic chickens, when I was a postdoctoral researcher at the University of New Mexico in Albuquerque. Chickens have been domesticated for so long that their husbandry is well understood, so we could use many of the same techniques employed for studies of flies or beetles. We know how to feed and house them, and we know how big an area they like to have to thrive. In our research, we were interested in how the elaborate ornaments of the roosters, including their fleshy combs and wattles, had evolved, and whether the hens preferred characteristics that might indicate a male's health.

I loved working with the birds, and it gave me a fondness for chickens and their kin that I harbor to this day, but that experiment was a nightmare. We had to ensure which hens mated with which roosters, take their eggs, individually mark them with pens, and rear them in incubators so their rearing environment was as standardized as possible. We also had to keep track of how many eggs were produced, how many chicks hatched from each family group, and then measure various attributes of the chicks as they grew up, including the size of the roosters' combs, the length of various feathers, and the kinds of mates the hens preferred. A large sample size is essential because of the complex calculations one needs to perform, so we were rearing hundreds of chicks, each of which had to be tracked individually. It was a gigantic exercise in bookkeeping, and I developed enormous respect for accountants who work with large data sets of any kind. Not to mention for poultry farmers.

It turned out that heritability was highest in rooster traits such as tail feather length or the color of the neck feathers, neither of which were all that important to the hens in choosing a mate. Intriguingly, males with larger combs sired larger chicks than males with scrawnier ornamentation, which is consistent with the idea that the comb shows a rooster's general vigor. Much remains to be done to pin down exactly how the attractive traits, and the preference for them, are inherited. It just won't be done by me.

Be that as it may, we, and many other scientists, have calculated heritabilities of behaviors ranging from courtship frequency in fruit flies and junglefowl to preening in Japanese Quail to learning in pigs. One paper[10] reviewed fifty-seven of such studies and found that the average heritability was 38 percent, though the range was substantial. This figure is well in keeping with the heritabilities of human behaviors and behavioral disorders such as anxiety disorders or major depression, though it is somewhat lower than the estimates of 50 percent to 60 percent for alcoholism and somewhat higher than the figures of 15 percent to 20 percent for extraversion and assertiveness. In a 2006 paper from the *American Journal of Psychiatry*, the authors conclude, "With respect to the broad patterns of genetic influences on behavior, *Homo sapiens* appears to be typical of other animal species."[11]

Two important conclusions follow from this kind of research. The first is that these numbers do not mean what many people think they mean. Heritability is not a measure of "how genetically based" a particular characteristic is, whether we are talking about a behavior like cheerfulness or a more objective trait like height. A heritability of 38 percent does not mean that a fly gets 38 percent of its courtship behavior from its genes and the remaining 62 percent from its environment any more than one gets 60 percent likelihood of becoming an alcoholic from one's parents. Remember, heritability only means something in reference to a population in the environment in which the characteristic was measured.

Like the zombie idea of nature versus nurture, however, heritability has become something of a zombie measurement, so that no matter how many times its limitations are explained, the misunderstanding revives itself. A case in point is the interpretation of a 2015 paper[12] that examined hundreds of human twin studies for traits ranging from developmental diseases to personality components to (quite intriguing to my mind) "Mental and Behavioural Disorders Due to Use of Tobacco." The authors noted that "across all traits the reported heritability is 49%." Media coverage and commentary on the work used that 49 percent statistic to conclude that nature and nurture were "tied," as if a longstanding battle had finally been resolved.

Not so fast. For one thing, the authors of the paper obviously could only review studies of traits for which there are data—if someone didn't look for the heritability of a trait, that trait didn't appear in the analysis. For another, as noted earlier, all the heritability estimates apply only to a particular place and time for a given population. IQ scores, for example, have famously increased in the

United States from an estimated seventy to eighty in the 1940s to closer to one hundred in the 1990s—and even ignoring the many, many problems associated with using IQ tests as a measurement, no one would argue that the human genes associated with whatever it is that IQ measures have changed that quickly. Instead, the environment changed, and the scores are only relevant (again, to whatever extent they ever are relevant) in that environment.

The second conclusion is that although we are fascinated by the idea that aspects of our own behavior are influenced by our genes in the context of our environment, whether a tendency toward optimism or our economic achievement, there is nothing special about the way the environment influences human behavior. All the caveats about heritability and its dependence on the environment, its nature as a population measure and not an individual one, and its lack of correspondence with "how genetic" something is, also apply to animals. Animals are not more or less controlled by their genes than humans, because neither of us is "controlled" by our genes at all. What that means is that while understanding the relationship between genes and behavior is valuable, humans do not have a premium on complicated interactions between genes and the environment. Eric Turkheimer,[13] whose observations about genetics I mentioned in the introduction of this book, proposed the "first law of behavioral genetics": all human behavioral traits are heritable. Turkheimer is, of course, a psychologist trained in behavioral genetics, so perhaps the stipulation of humans is understandable. But, in fact, behavioral traits are heritable whether they are in people or worms, and we shouldn't be more surprised about that finding in one or the other. Both humans and animals also influence their environment in many ways depending on their genes, as evidenced by the concept of niche construction discussed earlier in this chapter.

A slightly silly but frequently used example may also illustrate the way that heritability cannot be equated with "genetic." I am completely confident that every person reading this paragraph has a head. Presumably we can all agree that one is born with one's head—it is not a manifestation of the environment in which one lives, but a complex result of biological processes that occur during the development of an embryo. Although no "head-producing gene" exists per se, the existence of our heads depends on our genes. So what is the heritability of having a head? Zero. That is because any variation in head possession can be attributed to environmental factors, such as the preponderance of guillotines.

Despite all the limitations to its interpretation, understanding heritability is crucial if you want to examine how behavior can evolve. Recall that evolution means that genes are changing in a population over time in response to selection, so that, as in my earlier example, the birds became greener because the more camouflaged individuals had more babies. How much greener they get, and how fast, depends on two things: the strength of selection and heritability. Strength of selection means how much of an advantage a given difference in a trait confers; if only the birds that were slightly greener survived, while all their less-green compatriots were snatched up by sharp-eyed hawks, selection is stronger than if green feathers meant just a slight edge in camouflage. The greater the strength of selection, and the higher the heritability, meaning the higher the proportion of the variation in the trait is ascribed to genes, the more the population can evolve. So it's helpful to understand how genes influence behavior. But we won't get anywhere by pitting genes against the environment and expecting a winner to emerge.

Genes, Income, and Flies

None of the studies of heritability, whether on human twins or cattle, claim to identify particular genes or even groups of genes that are linked to a particular behavior—all they can do is evaluate the proportion of the variation we see that can be attributed to genetic variation, with "genetic variation" as a one-size-fits-all description. Over the last decade or so, however, advances in molecular biology, and the ability to sequence the genome, have allowed scientists to look much more closely at how people or animals with different characteristics differ at particular parts of their genome. It is this latter development that has led to headlines about genes linked to everything from sexual orientation to liking dogs to a predilection for getting divorced. In addition, the rise in genetic ancestry tests such as 23andMe has contributed to the idea that we can survey our genomes and pick out the parts that make us, say, athletic or slothful.

As many people have pointed out, that conclusion—and the idea that we can identify something like a "gene for sexual orientation"— is false. To understand why, it is useful to know just what is being compared in all those studies that garner "Genes Explain Your Income Level" types of headlines. The research uses Genome Wide Association Studies, usually abbreviated as GWAS, and pronounced "gee-wahss." To perform a GWAS, you first identify a characteristic of interest. In the earlier days of the technique, these were usually diseases, and medical applications are probably still the most common reasons for this research. Then you find a group of people with the disease and a group without it, and you examine some portion of their genomes—no one is actually sequencing all thirty thousand genes in the human genome—and look for differences in those small chunks. If the sample is large enough, the idea is that any consistent differences you see between the two groups are due to the difference in the presence of the disease. One can also use the results to develop a polygenic score or summary of the gene variants associated with the trait of interest.

These surveys are very useful, but they do not tell us "how genetic" any behavior, disease, or other characteristic might be. As I have already pointed out, genes do not code for behaviors directly, and they do not exist in a vacuum.

One of the most sensitive areas for exploration of this issue has been with intelligence, or at least with scores on IQ tests. Intelligence was one of the first behavioral characteristics that early geneticists wanted to understand, and the idea that we cannot change people's intelligence because they inherit it as a fixed, immutable property has dogged social programs for over a century. A set of articles in the *Wall Street Journal* in 2020 revisited this issue after a study was published that examined the association between genetic variation and some cognitive traits like how much education people had and their IQ. After seeing how much their work had been distorted in the media, authors Michelle Meyer, Patrick Turley, and Daniel Benjamin said:[14]

> IQ is not a fixed attribute of individuals and can be affected— for better and worse—by the environment in myriad ways. For example, in a society where people of color are denied access to childhood enrichment programs or adequate nutrition, a polygenic score for IQ might reflect genetic variants associated with skin pigmentation. Relatedly, in a sexist society, variants on the X and Y chromosomes, which determine biological sex, might be related to a variety of socio-economic phenotypes. Such polygenic scores would indeed moderately predict the IQ of people on average, but—and this is key—much of that predictive power would simply reflect social choices, not innate or immutable biology.

Everything that Meyer and her colleagues say here is true, but what many people do not appreciate is that it is true not only for people, with our rich social environments and complex development, but for animals. The problem isn't that GWAS or heritability are unsuitable for explaining how human behaviors are fixed by the genes, it's that this isn't what they explain at all, in humans or anything else. A recent study[15] of that tiny genetic powerhouse the fruit fly found that in a wide variety of behaviors, especially mating behavior, hundreds and often thousands of genetic variants were important. The research pointed to some interesting prospects for future work—for example, some of the genetic variation associated with body size was also important in how much male fruit flies court females. This raises questions like: What are the shared neurological pathways that cause such a relationship? In other words, we aren't going to uncover a gene, or even a handful of genes, that by themselves determine come-hither signals in fruit flies. This reality I hope underscores the futility of thinking we can ever do something comparable for income levels or divorce in human beings.

The Essence, or Instinct, of It All

Despite our efforts to eradicate the zombie of nature versus nurture and related misconceptions about what heritability really means, both are remarkably persistent ideas. In a survey published in 2010 of 1,200 American adults, 76 percent of respondents believed that "single genes directly control specific human behaviors."[16] I realize that scholars have long been lamenting the ignorance of the public on issues ranging from the geographic location of countries to the efficacy of antibiotics for treating viral infections, and that such surveys can be questioned for their propensity to generalize. With that caveat, this particular misconception is still troubling for at least two reasons. First, it means that the influence of the environment on all characteristics, whether behavior or not, is ignored, when in truth genes don't single-handedly determine anything. Second, and perhaps of even more concern, this misconception can lead to acceptance of bad, or even criminal, behavior. Psychologist Steven Heine from the University of British Columbia notes[17] that "men show increased moral acceptance of undesirable behaviors such as date rape when genes are even remotely implicated as opposed to societal forces."

So why can't we accept the limitations of these measures? Perhaps, as Heine and other psychologists, particularly Ilan Dar-Nimrod at the University of Sydney, have proposed, we are genetic essentialists. According to this view, many people harbor, as one of Heine's papers[18] says, an "innate set of psychological intuitions that lead us to think about genetic concepts in a highly inaccurate and biased way." Leaving aside the presumably unintentional irony of referring to misconceptions about genetics as "innate," essentialism can be a way to view all things as having an internal, immutable "essence" that makes them the way they are. This idea stems from Aristotle, who, as Heine and colleagues say, "famously proposed that every entity possesses an essence that ultimately makes it what it is and that, without such an essence, the entity would no longer be itself." Hence, as I mentioned in an earlier chapter, an ant has an "ant-like" essence that makes it an ant, apart from the more tangible things like its antennae, its small size, or its love of sugar and picnics.

Applied to genetics, this means that we imagine our genes as surrogates for that essence, a way for us to view our identities. Genes, like essences, are therefore seen as fixed, and so determine who we are. That attitude in turn might make us less inclined to change ourselves, or to attribute poor outcomes to irresistible forces, as with the date rape example. Research also shows that people ate more cookies after reading about potential genetic contributions to obesity than if they had read about the way the environment affects obesity. Both readings were accurate, but apparently being reminded of the way that genes play a role in our lives changed the subjects' behavior in the short term. A similar view has also been applied to human racial categories, with racial essentialists arguing that people of different races have physical and psychological differences that are both fixed by their genes and similar among all the members of a race. Neither of these generalizations are true, but the errors are, again, remarkably persistent, to the point where educator Brian Donovan suggests,[19] "There appears to be a hidden racial curriculum in biology textbooks that is learned by students but never purposefully taught by teachers."

Psychologists have explored the significance of essentialist views, and the difficulty they pose for social change, but I want to make a slightly different point here. It's true that genes don't give humans their "essence," and indeed the idea of ineffable essences is an odd holdover from ancient times. But here's the thing: animals don't have essences either, and their genes don't do any better at determining their identities. Many of the authors who provide caveats to the notion that genes determine behavior take pains to confine themselves to humans, pointing out that humans are incredibly complex. One such paper states,[20] "A typical human behavioral trait is associated with very many genetic variants, each of which accounts for a very small percentage of the behavioral variability." That is absolutely true. But it is also true for typical animal behavioral traits. Part of why the work on the *yellow* fruit fly mutants, or the ILP_2 gene, is so fascinating is that such relatively simple relationships are so rare.

A stand-in for that amorphous essence that governs what we do is sometimes called instinct. The concept of instincts is also quite old, and it is sometimes used to distinguish animals from humans, with the former supposedly acting instinctively, meaning without any cognitive process, and the latter able to reason and choose. People are sometimes surprised when I tell them that biologists don't use the word *instinct* much in animal behavior research anymore. To me it is a non-explanation; saying that a bird building a nest is behaving instinctively, as the *Star Tribune* newspaper article I mentioned earlier did, just fobs off the question. All the word *instinct* tells us is that whoever called the behavior an instinct didn't know—and indeed, we rarely do—how the environment and genes interacted to produce that particular behavior. Instincts don't exist, the same way that essences don't exist—animals aren't born carrying a mating instinct, a feeding instinct, or any other kind of instinct.

Recognizing, once again, that neither humans nor behaviors are special cases frees us from having to explain their exceptionalism. More important, it means that we can see the extraordinary flexibility and complexity of nonhuman animals and most if not all of their characteristics, whether physical or behavioral or an inextricable mix of both.

4

Raised by Wolves— Would It Really Be So Bad?:

The First Domestication

D
r. Stephen Lea is my candidate for bravest man in the world. An emeritus professor of psychology at the University of Exeter in England, he published a paper in 2018 with Britta Osthaus[1] titled, "In What Sense Are Dogs Special?" In this study, which I explain in detail later in the chapter, the researchers concluded that although dogs can perform some cognitive tasks perfectly well, they are not, as the title says, particularly "special."

The reception to their work was not appreciative. "Your Dog Is Probably Dumber Than You Think, a New Study Says," smirked a typical headline from *Time* magazine. "Shocking News for Dog Lovers: Canines Aren't as Smart as We Think" bemoaned the *Mercury News*. One article with the headline "New Research Suggests Dogs Aren't Exceptionally Smart" has a quotation from Lea as a subheading of sorts: "We're certainly not saying that dogs are dumb," which bespeaks a certain desperation. Lea also tried to pacify the dog people in another interview with "Dog cognition may not be exceptional, but dogs are certainly exceptional cognitive research subjects."[2] No one seemed placated by these qualifications. The paper by Lea and Osthaus could have been titled "Goats, Pigs, and Pigeons Are Special, Just Like Dogs," but that lacks a certain zing.

Never mind that the study didn't show that dogs were stupid. But it turns out that saying anything negative about our best friends is tantamount to insulting God, mother, and country all at the same time. Why is that? Why are we so obsessed with all matters canine, from the evolutionary origin of dogs to the way they can best be trained to stay off the sofa? What really does make for a good dog?

From my perspective in this book, the domestication of animals, and of dogs in particular, is one of the best ways to understand how behavior evolves, and how it is linked to physical characteristics like size and shape. Domestication has been a cornerstone for many scientific ideas over the centuries. Charles Darwin relied on domestication to formulate his ideas about evolution. He was particularly engaged by pigeons, which were being bred by British Victorians to have many elaborate feather colors and patterns. The birds helped him argue that even the most outlandish types of pigeons arose from a single species, the humble Rock Dove. Science journalist Richard Francis points out in his book *Domesticated* that "human civilization as we know it would not exist"[3] without domesticated plants and animals, since they enabled us to leave lives of hunting and gathering to settle down. Similarly, other scientists have made sweeping claims about the impact of domestic species, dogs in particular. Certainly life as we know it would be vastly different without our pets, crops, and meat- or milk-yielding beasts. What is more, when it comes to domesticated animals, the hallmark of domestication is behavior, not appearance. Cows that gave more milk, or sheep that had thicker wool, would be of little use if they never stood still long enough for us to herd them into a barn.

But of course there is more to our interest in domestication than just the practicality of keeping animals. We obviously love our domestic animals, or at least highly value them, in the case of the less-cuddly ones like camels or reindeer. At the same time, they showcase our dilemma about the fixed versus malleable nature of behavior. If, on the one hand, instincts are immutable, how did early humans make a lap-sitter out of a fierce predator? But if behaviors are simply induced by the environment, well, a similar question comes to the fore: How could those humans have taken placid canine parents and counted on them to produce placid puppies?

Our understanding about the domestication of dogs has expanded enormously over the last few years, as scientists sample ancient DNA and use sophisticated statistical techniques to evaluate the similarity between modern breeds and dogs from thousands of years ago. You can now use ancestry websites to examine not only where your dog came from, but also its likely personality characteristics and whether you should do things like enroll it in agility training. But it is turning out that the simple story of a wolf that lurked at ancient campsites and then stayed because of mutual convenience is much more complicated than it first appeared. Virtually everything about what we thought we understood about dogs is now up for grabs—their origin, the way they were domesticated, their behavioral abilities and empathetic connection to humans, and even whether they are distinct from their wolf ancestors.

The Wolf That Came in from the Cold—or Didn't

Where did dogs come from? The simplest answer to that question is wolves: domestic dogs share much of their genetic material with wolves, they look like wolves, and archaeological evidence suggests that dogs started out in places where wolves lived. After that, however, the story gets complicated quickly. Estimates of the time of the first dogs range from as many as 135,000 to as few as 10,000 years ago, an astonishingly large spread for an animal that has been so intensively studied.

Part of the ambiguity stems from what counts as evidence of a dog rather than of a wolf, which then leads to the equally thorny question of how the shift from wild creature to tame companion occurred. A 26,000-year-old mark in Chauvet Cave in southern France seems to show, next to the footprints of a young boy, the

paw marks of—well, either a wolf or a large dog.[4] But which? Other
scientists looked at fossil bones associated with human encampments and suggested that something like a large dog may have
been present in Europe up to 36,000 years ago.[5]

Fossil remains, however, are frustratingly limited in what they
can tell us about the relationships between wolves and dogs, or
about how various dog breeds themselves are connected. One has
to infer a great deal simply from measuring the bones themselves.
Because individuals differ in any species, generalizing from one
bone from one animal to an entire species can be difficult. Just in
the last few decades, scientists have been able to rely instead on
DNA, either from modern dogs to determine how much they differ
from each other and from modern wolves, or from ancient samples
that come from bones excavated at archaeological sites around the
world. Such studies have variously concluded that dogs originated
in the Middle East, in southern China, and Europe.

A study published in 2013 by Robert Wayne, an evolutionary biologist at UCLA specializing in dog genetics, and his colleagues compared DNA from fossilized bones, from modern-day wolves, from
several dog breeds, and from coyotes—using the latter as what is
referred to as an "outgroup," a more distant relative that serves as
a benchmark for the other comparisons.[6] Their analysis yielded an
estimate of dog origins in Europe from 18,000 to 30,000 years ago.
The scientists imagined that as wolves interacted with humans,
they migrated along with them, thus becoming separated from
other wolves and subject to humans seeking and selecting the more
sociable associates. This research also cast doubt on the suggestion
that dogs arose in East Asia, because those studies, Wayne and his
colleagues attest, used DNA not from dogs but from modern-day
hybrids between wolves and dogs.

But other scientists, most notably Peter Savolainen at the Royal
Institute of Technology in Stockholm, weren't convinced. They
pointed out[7] that Wayne's samples hadn't included enough individuals from Asia in the first place; you can't find something from a

place where you haven't looked for it. A rather contentious debate ensued, with arguments about who was sampling appropriately, who was using the right DNA, and what parts of the fossils to measure. Paleontologist Mietje Germonpré measured a 32,000-year-old skull from a museum in Brussels and identified it as a dog, which would have put the origin of the domestic animal rather early, but her work too was hotly contested. In an article in the journal *Science*, she is quoted[8] as saying, "It's a very combative field. More than any other subject in prehistory." In case you have never met any archaeologists, let me point out that this is really saying something.

A new slant on the story appeared in 2020,[9] when a group of scientists published the results of a study of sled dogs, those hardy workaholics of the Far North. Sled dogs seem like a niche breed, but they have ancient origins; carved bone and ivory tools resembling dog harness fasteners have been discovered at sites in Siberia dating back nearly 10,000 years. The researchers compared the DNA from ten sled dogs from Greenland, a bone from a 9,500-year-old Siberian dog that was associated with evidence of sleds, and a wolf bone also from Siberia that was dated to about 33,000 years ago. The ancient dog bone was similar to modern sled dogs and to the ancient wolf, but not to modern American wolves. This suggests that the sled dogs in the Iditarod can trace their ancestry back to Siberia, and that they are not a modern breed originating only a few centuries ago, as one might have thought given their current use.

Perhaps even more intriguing, the genetic similarities between old and new sled dogs suggested just how selection for domestication changes the genome. Most domestic dogs have multiple copies of a gene that is useful in digesting starches, widely viewed as a way that dogs became better able to live with humans and our carb-reliant diets. Even among humans, populations such as those from East Asia, where rice has been a staple for many thousands of years, have more copies of the gene for starch digestion than those populations from places where starches are not as major a food source.[10] Indigenous populations in the Arctic fall into the latter category, and, no surprise, so do the sled dogs. The sled dogs do have genes that help them digest the fatty acids that are prevalent in a diet heavy in well-larded meat. One of those genes is similar to a gene found in polar bears, though of course the bears and dogs have not shared a common ancestor for a very long time, illustrating the way that evolution uses different solutions to solve the same problem.

How Did Dogs Become Dogs?

The ability of dogs (at least in certain areas) but not wolves to digest starches leads to another hotly contested place that is relevant to dog evolution: the garbage dump. The conventional wisdom is that tens of thousands of years ago, humans and wolves were bitter enemies. As Richard Francis puts it, "All wolf-human interactions were overtly hostile. We competed fiercely for the same prey and probably killed each other at every opportunity."[11]

Then, the story continues, some wolves started hanging around human encampments and scavenging food from the trash heaps. The wolves that were less aggressive toward people, perhaps bringing their pups with them, gained by being fed. The humans— though the literature is less clear here—initially simply tolerated their presence and then later favored some individual wolves and even took in their puppies, raising them as pets. This scenario is suggested to have taken place sometime around the beginning of agriculture, perhaps 10,000 years ago. After many generations of people selecting the tamest wolves, the animals changed not only in their aggressiveness but also in their appearance.

Alternatively, another theory suggests that the wolves domesticated themselves. Again, the friendlier individuals were spending time near humans, although no one deliberately tried to raise puppies or even encourage interactions. Eventually, however, the self-domestication meant that ever-tamer wolves gained more advantages from humans, and dogs were the result.

Both of these pathways take for granted the mutual enmity expressed by Francis, and the active role played by humans, and humans alone, in molding the friendly attributes of dogs out of the vicious clay of wolves. But what if those assumptions are incorrect? Ray Pierotti, a biologist at the University of Kansas, and his colleague Brandy Fogg believe that the image of the Big Bad Wolf is, as the name suggests, a fairy tale, and a Eurocentric one at that. In their book *The First Domestication*,[12] they propose a more cooperative origin of dogs. Drawing from indigenous accounts, mainly from North America, Pierotti and Fogg argue that many tribal groups cooperated with wolves as equals, hunting together and viewing the predators as colleagues, not enemies. They contrast the language used to refer to wolves by Western scholars—with descriptors such as "ferocious"—with the stories by indigenous people, which mention "companion" or "grandfather." This appreciation of wolves suggests to the scientists that wolves and humans evolved together. They cast doubt on the "dumpster" origin story, pointing out that wolves' outstanding hunting ability would make it unnecessary for them to seek out human refuse to eat.

Instead, Pierotti and Fogg speculate that perhaps a solitary wolf that was driven from its pack joined the only other available hunters in the area, namely people; if those hunters were willing to partner with the wolf, both parties could benefit. The wolf-assisted humans could then expand into parts of the world that did not previously have wolves, and the appearance and behavior of wolves could have diverged into a more doglike form. This set of events could be repeated in various parts of the world, and we could see the emergence of dogs without having to posit a relationship in which humans did all the manipulating and wolves were simply receptacles of selection.

Other scientists are also rethinking the story, with an eye toward mutual hunting benefits rather than exploitation. It has even been suggested that wolves helped early humans extend their range by enabling them to bring down large prey. And anthropologist Pat

Shipman argues[13] that people worked with the earliest forms of dogs to hunt large numbers of mammoths, beasts that would have been hard for humans to kill with the tools they had at the time. The genetic and fossil evidence about dog domestication is still being gathered, with the boundary between wild and domestic still blurred.

My, What a Big Brain You Have

Pierotti and Fogg also maintain that wolves and dogs are not nearly as different as people like to think. In central Asia, for example, you would be hard-pressed to tell dogs from wolves, at least from their outward appearance. And archaeological remains can be tricky to interpret; what looks like a wolf to one can be a dog to another. To Pierotti and Fogg, all dogs are wolves, but not all wolves are dogs, meaning that the boundaries between the two are not fixed; a poodle is clearly a modern invention (most of the newer dog breeds only date back a couple of hundred years at most), but they are still sufficiently enfolded within the Venn diagram circle of wolves to make a hard-and-fast distinction questionable.

The authors suggest that as humans and wolves cooperated in hunting, wolves first changed their behavior to become more dog-like, in a way that left the appearance of the two animals indistinguishable. Indeed, the evolution went both ways, with human behavior changing even as there was little effect on human physical characteristics beyond, intriguingly, a reduction in the skeletal shape and size differences between men and women. According to this view, wolves made early humans more cooperative in a way that distinguishes us from our closest evolutionary relatives, the chimpanzees. Their position is not widely shared; humans seem to have evolved to be more pacific than chimps for a variety of, again hotly contested, reasons, including the notion that humans themselves are self-domesticated.

Complicating this issue are popular, rather than scientific, conceptions about just how similar wolves and dogs really are. On the one hand, we breed dogs like bichon frises or Pomeranians, in which anybody would be hard-pressed to see a resemblance to a wolf. On the other hand, we (and the pet food companies, or at least their advertising writers) seem to be convinced that dogs are still, at heart, wolves—witness the response to an article in the *New York Times* suggesting vegan diets for dogs. The article noted that in part because of the aforementioned changes in genes allowing for the easier digestion of starch, it is possible to give dogs a plant-based diet, although experts disagree about whether a dog that eats a vegan diet will get all of its nutritional requirements for long-term health. Be that as it may, readers had strong opinions.[14] One suggested taking a vegan dog to a sheep farm and then "watch how fast the 'vegan dog' becomes a wolf." Concern about "forcing dogs to eat a diet they were not designed for" was met with the rejoinder that if the mere ability to consume flesh were the only consideration, humans should be eating each other, since cannibalism has been seen in many societies. Another flatly stated, "Dogs are not meant to be vegetarians. If you can't deal with that, get yourself a goat instead." Running through these reactions about inner wolves is the idea that animals—and perhaps people—have an essential nature—in this case, an internal wolfness, so to speak, that is dictated by their genes. That essentialism, as I noted in the previous chapter, will always lead us astray.

At the same time, we eagerly look to studies that show us the ways that dogs and wolves differ in their behavior. For example, a widely cited study from 2017[15] examined the ways that dogs versus wolves cooperated in a task at the Wolf Science Center in Vienna, Austria. Pairs of both species—raised in as similar a way as the investigators could manage—were presented with a tray of food on the other side of their cage. Two ends of a cord were threaded through rings on the tray such that if both were pulled, one by each member of a pair, they get the snack. But if only one did so, there were no treats for either. Though neither canid did spectacularly at this test, the wolves surpassed man's best friend by a wide margin, with five of seven pairs managing to acquire the tray on at least some of their attempts, while only one of the eight dog pairs ever figured it out.

Much was made of this difference, with the experimenters concluding that it showed how dogs are not natural cooperators, unlike the pack-dwelling wolves. Perhaps, the researchers speculated, dogs lost their ability to work in a team as they forged their bonds with humans; you don't need to collaborate to find a discarded carcass at a refuse site the way you do when you attempt to take down a moose. On the other hand, Laurie Santos, a cognitive scientist from Yale University, suggested that the dogs are perfectly good at cooperating, but they perceive the physics of the problem differently, and so can't solve the problem.[16]

If the whole let's-pull-on-a-string-together test sounds too much like some kind of team-building exercise from a corporate retreat, how about a nice game of fetch? A 2020 paper[17] by Christina Hansen Wheat and Hans Temrin from Stockholm University determined how eight-week-old hand-raised wolf puppies responded to investigators—charmingly called "puppy assessors"—throwing a tennis ball. First, the puppy had a chance to play with the ball on its own, and then the assessor "encouraged" the puppy to retrieve the ball.

The results were, again, not stunning; just three of the thirteen wolves brought back the ball even some of the time, and only one, named Sting, managed to do so on all three of the trials. And an eminent researcher in dog genetics, Elaine Ostrander, pointed out that it's not clear what the test revealed, since many dogs won't retrieve balls either. Nevertheless, this accomplishment was lauded in the popular media as indicating an ancestral willingness to play with humans residing deep in the wolves' DNA. In turn, that suggested to some that you don't have to work that hard to domesticate a wolf—they have had it in them to be house pets all along.

Except, of course, when they don't. Although Pierotti and Fogg point out that many of the instances of vicious behavior by wolves kept as pets are actually cases of dogs that simply look like wolves, it's clear that, as yet another article about wolves and dogs in the *New York Times* put it, "No matter how you raise a wolf, you can't turn it into a dog." Or as Patricia McConnell, a noted animal behaviorist who consults on dog behavior, wrote: "Dogs aren't wolves, pure and simple. Except, uh, they are. Sort of. Sometimes." She goes on to explain that while we all see the similarities in behavior, and we know that dogs evolved from wolves, wolves can rarely be housebroken, and they use different cues to assess friendliness or aggression than do dogs.[18]

She also debunks the notion that wolves live in strict social hierarchies, with a single alpha pair, and that the dominance relationships in a pack are carried over into dogs, with humans the "alpha" individuals. First of all, social status in animals is complicated, particularly for big-brained species like wolves. Dominant individuals don't always get their way, and different wolves can approach being alpha differently. And even in feral packs, dogs do not replicate the social system of wolves, which means that dog trainers who urge owners to constantly assert their dominance, as though life with a terrier in the suburbs replicated the *Call of the Wild*, are, well, barking up the wrong tree.

What this means is that, while wolves can certainly come to associate with humans without eating them (a basic requirement for domestication in anyone's book), they are not simply dogs that haven't been given enough attention while young. A great deal of research on wild wolves has shown that they have markedly different behavior from dogs at the get-go, including care of the young by both parents, not just the mother, as is the case for dogs. Though both use their noses in ways that make humans seem effectively scent-blind, wolves and dogs emphasize smell differently, with wolves using it as their exclusive way of interacting with the world as pups, and dogs relying on a combination of sight, smell, and sound. As I detail later in this chapter, many researchers are using the latest tools in genetics to better understand the ways that genes contribute to the changes we see in domestic dogs compared with their wolf ancestors.

Another much-touted difference between wolves and dogs is a dog's ability to recognize when a human points to something. If you present the animal with two bowls of food whose contents aren't visible to it, and point to the bowl with the snack, will the animal go where it's directed? Again, sometimes. Dogs generally do better than either wolves or chimpanzees at this task, particularly when they are young (and are on a par with two-year-old children), although if the wolves have spent a lot of time with people they catch on more quickly. A key to success seems to be the ability of the animal to look the demonstrator in the eye, something wolves are simply not as prone to do. The results have caused researchers to wonder about whether dogs and wolves have similar abilities, but the ages at which they take hold are different in the two animals.

I am intrigued by the idea that cultural biases have shaped the way we view domestication, and I think that vilification of wolves has probably clouded our opinions about how dogs evolved. But I am not convinced that it is only our Eurocentrism that makes us hold wolves apart. Instead, the dog-wolf divide illustrates yet again just how intertwined genes and the environment are in producing behavior. Yes, you can get a wolf puppy to play fetch, and be petted, and walk on a leash, so wolves obviously possess whatever genetic material is required for such interactions. But you have to live with those puppies 24/7 for several weeks before that becomes feasible, whereas a dog will oblige you with affection with a fraction of that type of contact. At the same time, you cannot simply grab a feral stray dog off the street and expect it to sit, stay, or even refrain from biting you. And the ability to play with an object, or rescue people in distress, varies widely across breeds, and individuals, and when and where it happens. These behaviors are more likely to occur in dogs than wolves, but they are not universal. This variability also casts some doubt on the generalization about one species or the other from a round of tests using just a dozen or so individuals.

The thing is, regardless of whether you are talking about a dog or a wolf, there's no such thing as an innate tendency to fetch a ball, or a stick, or to pull on a string to get a tray of food. The differences between dogs and wolves illustrates the point I have been making all along: behavior isn't something that is unilaterally drawn from a gene or group of genes, and it isn't something that emerges from an experience during puppyhood. Instead, it's both.

Domesticated in Life, Fur, and Paws

This combination of genetic change and early experience is part of the complex story of how domestication shapes behavior and appearance at the same time, but still can't alter everything about an animal. Not only can't you make a silk purse out of a sow's ear, you can't really change a sow's ear much at all, at least without some unforeseen consequences.

To understand this, we need to delve deeper into what domestication really means. Many people use the word as though it is synonymous with tameness, or lack of aggression, hence the aforementioned tendency of dogs to be friendly and wolves not so much. Anthropologist Richard Wrangham points out[19] that aggression comes in two basic flavors, reactive and proactive, a distinction first made by psychologist Leonard Berkowitz in 1993. Reactive aggression is the kind that occurs when a wolf snaps at a rival during a fight, or a child hits a classmate who snatched a desired toy. It is immediate, emotional. Proactive aggression is planned, the kind manifested by premeditated murderers. Wrangham argues that the two have different functions and are subject to different selective forces, as I will explain in the next chapter. Tameness, then, is related to lower reactive aggression, and it seems certain that people would have bred, or at least wanted to associate with, the wolves with the lowest levels of such aggression.

But domestication is more than tameness. An individual animal can be tame, because it has become accustomed to being handled by humans. Species differ in the degree to which they will respond to such treatment. People can approach the hand-reared wolves I mentioned earlier, and these wolves will even show them affection. But they are still unpredictable in their behavior. It took many generations of selective breeding, in which the friendliest wolves had puppies, and then the friendliest of their puppies were chosen to breed, and so on, until dogs came to be reliably tame.

So are dogs exactly like wolves, except with more docile demeanors? We all know the answer is no; dogs differ from wolves in many ways, both physically and in their behavior, and the links between the two provide some of the strongest evidence for my contention that both evolve in concert. Charles Darwin, in his work on domestication,[20] was one of the first scientists to note that in addition to being less aggressive, domesticated mammals such as dogs also have floppy ears, white spots in their fur, curly tails, and white feet. At least a subset of these characteristics also appears in domesticated cows, horses, and cats when compared with their wild forebearers. Domesticated animals also tend to be smaller, with shortened faces and relatively smaller brains, than their wild cousins.

This constellation of characteristics is called the domestication syndrome, and its significance has been puzzled over since Darwin. Why should a white tail have anything to do with aggression, or with a smaller jaw? One explanation is that all the traits evolved as the various species lived with humans and they did not have to survive on their own in the wild. Hence white patches were seen as attractive by humans, or were useful in telling animals apart. Floppy ears were not so much selected for as not selected against—if floppy ears make an animal's hearing less acute, that wouldn't matter much if that animal is living a sheltered life and doesn't need to be ever vigilant for the sound of a predator. Modern breeds, of course, do not all share these signatures of domestication, but those more individual departures from the syndrome might have come later.

The problem is that this seems like special pleading for what is really a rather motley collection of attributes. If pressed, one can come up with an evolutionary explanation for almost anything, and these just-so stories, as these explanations are called, after the Rudyard Kipling tales, often have little behind them other than a surface plausibility.

To consider an alternative, we need to turn away from dogs for the moment. We also have to travel, mentally at least, to Siberia, and to a story that reads as much like a Cold War spy novel as it does a science experiment. In 1959, a Russian scientist named Dmitri Belyaev decided to try and untangle the puzzle of how dogs became domesticated by taking the then unheard-of tack of replicating the domestication process in real time.[21] He and his colleagues took silver foxes, widely bred in vast Siberian farms for their luxurious pelts, and worked to turn them into friendly house pets. It was a deceptively simple process: take the puppies from only the friendliest foxes, breed them, and repeat. It only took ten generations to get foxes that licked the faces of their handlers, something that the original foxes would never do. What is more, selecting for tameness and tameness alone also produced genetic changes in all kinds of aspects of the foxes' appearance: fur color, ear shape, and multiple breeding seasons per year, instead of the usual one. Belyaev and his colleagues argued that these changes were by-products of domestication, and part of the syndrome seen in other animals.

The experiment he started is still continuing, over sixty generations later, with biologist Lyudmila Trut in charge, and it is a landmark study in understanding domestication.[22] The spy novel part comes in because Belyaev began his project as the Soviet Union was shaking off the disastrous influence of Trofim Lysenko, the director of biology under Stalin. In keeping with the idea that the government could engineer a perfect society simply by manipulating the environment—whether of crops or people—Lysenko rejected the burgeoning field of genetics and promoted agricultural improvement through the inheritance of acquired characteristics, a la Jean-Baptiste Lamarck, an eighteenth-century naturalist. Lamarck is best known for proposing the idea that animals (and people) could permanently change their traits through sheer effort, with the classic example being giraffes straining to reach high branches and passing on their stretched necks to their young. Lysenko applied the same idea to growing crops, and his efforts resulted in the failure of Soviet agriculture. But for decades he was highly influential, and Belyaev and Trut had to pursue their research in the face of lingering skepticism at best and the threat of imprisonment at worst. They persisted, and the work has yielded much data about how behavior changes over time.

But having identified a constellation of domestication-associated characteristics in body as well as behavior, Belyaev didn't know how they became connected; genetics simply wasn't far enough along to allow investigation of the mechanism. In 2014, biologists Adam Wilkins and Tecumseh Fitch, along with Richard Wrangham, published a paper[23] suggesting that the domestication syndrome was due to changes in the migration of neural crest cells during embryonic development. Neural crest cells are part of the developing vertebrate embryo early on but disappear as the animal grows. The cells do not become any particular organ, but they influence the way that many different parts of the body develop, from tooth size to ear structure to the formation of the adrenal glands that produce hormones important in aggression.

This idea has been popular, and it does seem to provide a tidy theory uniting what had seemed to be a disparate set of characteristics. And Wrangham has extended it in his book *The Goodness Paradox* to suggest that the low reactive aggression is associated with the physiological traits of the domestication syndrome.

Recently, however, a snag in the story has emerged, with questions about the foxes used in the Russian study and hence the generality of the domestication syndrome itself. It started with a chance visit by the late biologist Raymond Coppinger to Prince Edward Island in Canada, which houses the International Fox Museum and Hall of Fame.[24] Coppinger, who had written extensively on dogs, saw photos at the museum that looked like the foxes in Belyaev's experiments, showing animals that had been captive-bred since the late 1800s. The foxes were already somewhat tame, with people walking them on leashes and holding them in their arms. These animals were used to establish the Russian fur farms in the 1920s. Belyaev himself had never claimed that he started with completely wild foxes, but he did refer to the starting population as "wild controls," which may have given the impression that his study demonstrated how one could go from wild to domestic after a relatively small number of generations of selection.

Spurred by Coppinger, a group of scientists led by Kathryn Lord from the University of Massachusetts and the Broad Institute in Boston traced the history of the Russian foxes and reexamined the domestication syndrome in a wide range of animals.[25] They analyzed the DNA of the experimental foxes and concluded that they derived from the Prince Edward Island population. Lord and her colleagues concur that Belyaev's experiment offers rich fodder for the study of how behavior is inherited, but suggest that the rapid response to selection occurred because the foxes were already on a path to tameness, and hence the evidence doesn't support Belyaev's rapid process of domestication. It is worth noting that foxes seem particularly prone to domestication, making it hard to draw conclusions about other animals based on them; for example, Native Americans brought descendants of grey foxes from the Northern Channel Islands off the coast of California to the Southern Channel Islands and kept them as pets.[26]

As for the suite of characteristics said to accompany domestication, Lord and her colleagues examined nine kinds of other domesticated animals using three criteria to see how well the syndrome held up.[27] First, the characteristic had to appear more or less at the same time as tameness during the course of domestication. Second, the characteristic needed to be more common in the selected population than in the population it came from. Third, the characteristic needed to be associated with tameness within an individual, so that the tamest dogs, or rabbits, or goats had to also be the ones most likely to show the white fur patches.

The result? None of the species fulfilled all the criteria, though each showed some of the predicted associations. Another study by a different group of researchers focused on dogs, and examined seventy-eight breeds for correlations among the different aspects of the domestication syndrome—were white feet found with curly tails? Such associations would be expected if the syndrome occurs because the same mechanism governs all the various components. Again, the answer was no, and the authors concluded that dogs, at least modern breeds, do not illustrate the syndrome very well.[28]

What does that mean about the domestication syndrome? I am sure we have not seen the last of this debate, but it is clear that while behavior and appearance can both respond together to selection, whether imposed by humans or by nature, the relationship between them is complicated. And that complicated interaction between genes and the environment is what we expect for all characteristics, not just in dogs or in the syndrome of traits that accompanies domestication.

Who's an Unexceptional Doggie? You're an Unexceptional Doggie!

Dogs can do some amazing things. So can wolves. And the two differ in ways that are interesting to consider in light of dog domestication. But what about the question posed by Lea and Osthaus that I used to start the chapter—are dogs special?

No question, people love their dogs, and have for, well, as long as people and dogs have been together. And we think the feeling is mutual. Charles Darwin, who used domestication to understand how selection by humans could be a model for selection by nature, mused: "It is scarcely possible to doubt that the love of man has become instinctive in the dog."[29] One could see this mutual love in the flood of COVID-19 pandemic puppies, as people in the United States emptied out shelters and breeders to cozy up with affectionate canine companions.

And for reasons that aren't clear to me, at least, people seem to have a stake in showing how their dogs, in particular, are geniuses. Dogs have been used in all kinds of cognitive studies, from the work of Pavlov on classical conditioning to tests of their ability to recognize a human in distress (spoiler: not very good). While some of that use of dogs is sheer convenience, I am willing to bet that there is more to it than just using dogs as a rat substitute.

One of the newest salvos in the how-gifted-are-dogs discussion has to do with speech. No one (yet) has actually taught a dog to talk, weird YouTube videos of dogs barking "The Star-Spangled Banner" notwithstanding. Still, people have long claimed that their dogs understand exactly what they say. To examine this claim, a group of Hungarian researchers set out to determine how dogs' brains respond to human speech.[30] They trained twelve pet dogs (border collies, retrievers, and a German shepherd) to lie down in an fMRI machine, and then had humans speak while the scientists recorded the responses in the dogs' brains. The humans repeated either positive words, like "clever" or neutral ones, like "if," with the same intonation.

The dogs responded differently to the two types of words, but the researchers hastened to say that this did not mean the animals understood the separate meanings of the words. Instead, they were interested in how the brains processed the sounds. The dogs seemed to use one part of their brain to respond to the emotional component of the words, and another part to respond to the word itself. Humans do a similar kind of partitioning when listening to meaningful versus meaningless sounds, which suggests that the ability to process vocal communication may be shared across a range of mammals.

Which brings us to the Lea and Osthaus study.[31] What the scientists did was straightforward. Dogs may be smart, but are they any smarter than you would expect? To answer that question, Lea and Osthaus picked three groups for comparison. First, they looked at studies of other species that are related to dogs evolutionarily—members of the group Carnivora, meaning meat eaters, including African wild dogs (which are not the same as feral or street dogs) and cats. Then, they considered dogs as social hunters. Other groups of social hunters include dolphins and chimpanzees. Finally, they examined horses and domestic pigeons, both of which are of course domesticated animals like the dog, and which are also amenable to a lot of training by humans. Each of these lenses provided a different but complementary way to think about dogs' abilities.

Any such comparison is bound to have an arbitrary element to it—one could argue for or against including any number of other species, from ferrets to elephants. But the idea was to see if the dogs performed in a way that was unexpected, given their place in any of the three grand schemes of animal life.

The result, as I've already intimated, was disappointing, at least if you are one of those my-dog-should-be-in-Mensa types. Yes, dogs do well at discriminating complex visual patterns, like telling human faces apart, but so do chimps and pigeons. Dogs are good at smells, but they are bested by pigs, which can even distinguish between the odors of familiar and unfamiliar people. Dogs are not particularly skilled at what Lea and Osthaus term "physical cognition"—recognizing the consequences of manipulating objects like strings attached to food, or understanding where a moving object will come to rest. Faint consolation may lie in knowing that cats do not do any better than dogs, though raccoons, unsurprisingly, excel at such tasks. Despite the heartwarming nature of movies like *Homeward Bound*, dogs aren't even particularly good at navigating over long distances.

None of this means, as both authors point out repeatedly, that dogs aren't wonderful companions. They can point (perhaps not literally, but that's okay) to the ways that different species have different abilities. Osthaus said in an email to me that by emphasizing other animals' skills, she was "doing my bit for animal welfare, really, as I pointed out pigs' amazing abilities, as well as pigeons'."[32]

A word about the bias of using human senses as a starting point is in order here. Many people have noted that so-called intelligence tests are biased toward the experimenter's worldview. We think animals are smarter the more they are like people, and the more they rely on human senses, primarily vision, rather than, say, scent, which plays such an important role in the lives of dogs. Science writer Ed Yong pondered the famous mirror test,[33] in which a spot is placed on an animal's body in a place where it is invisible unless the animal sees it when it looks in a mirror. Most animals do not recognize their images in a mirror, with chimps and elephants as noteworthy exceptions that reach for the place on the body that has been altered. The results are sometimes claimed to show that a species has a "sense of self" and can recognize the existence of its own body. But dogs are not particularly visual creatures, and rely instead on smell. If we could give them the equivalent test in odors, would they excel?

Perhaps. But a larger point—and one that was recognized in the work by Lea and Osthaus—is that it doesn't make sense to simply pick on an animal, no matter how beloved, and try to rank it according to a scale that only works in a single dimension or on human-centric traits. Yong quotes psychologist Alexandra Horowitz saying, "I think that all of these abilities, which we mostly decline to see in nonhuman animals until they've passed our tests, are defined in a far too binary way."[34] In other words, dogs are good at things that make sense for dogs to be good at, which is a rather unsatisfying answer but one that makes sense from the standpoint of evolution.

Dogs Rule with Their Drool

New techniques in genetics let us understand more and more about the link between how an animal looks and acts and its genes. What about combining the two? Elinor Karlsson is the head of a lab studying, among other things, dog genetics at the Broad Institute.[35] Kathryn Lord, who led the work casting doubt on the domestication syndrome, is part of her research group. Karlsson has an ambition that verges on the grandiose, and I say that with all the respect in the world: she wants to sample the DNA of every single dog in the United States, and perhaps elsewhere, and link those genetic signatures to details of behavior.

In a project called Darwin's Dogs,[36] she is, as the project's website says, "following the pawprints of evolution." More prosaically, the lab is enlisting dog owners who send in a sample of their pet's saliva (something dogs are more than willing to provide) and answer questions about the dog's behavior. Does Fido cross his paws before lying down? Does Princess like to fetch balls? Using sophisticated techniques in genetics, some of which I outlined in the previous chapter, Karlsson and her colleagues are primarily trying to understand behavioral and psychological disorders such as canine compulsive disorder, a syndrome I will discuss in more detail in chapter 6. But the project also provides a powerful tool for understanding the evolution of behavior more broadly.

The project has already pinpointed a gene that seems to be important in narcolepsy, the sleep disorder, and is examining genes that might predict a dog's success at becoming a working dog that can sniff out bombs or help people with disabilities. Despite many years of intensive training and breeding programs, the success of animals in the latter can be well below 50 percent, not the best return on investment. If Karlsson and her colleagues can understand the genetics behind the dogs that pass the test, that rate could be increased.

Other large-scale programs to understand dog behavior include Dognition,[37] a program run by Brian Hare at Duke University. "Find the Genius in Your Dog," its website proclaims. Owners perform a series of tasks like yawning in front of their dog to see if it yawns back (a test for empathy, or at least the strength of the bond between a dog and its owner), and send the results in to the site. In return, they receive a characterization like "charmer," "maverick," or "Einstein." The idea is not to rank your dog against others, but to appreciate its unique qualities.

Along with his colleague Vanessa Woods, Hare points out that the last decade has seen a proliferation of books about the inner lives of dogs,[38] as our interest in them has risen. Woods and Hare argue that cooperation between humans and dogs, as well as among humans, has been key to our mutual survival, and even our success at outcompeting the Neanderthals. Along with Wrangham, they see us humans having domesticated ourselves, becoming friendlier to one another in the process.

As someone who studies animal behavior, I applaud the increased recognition of the cognitive abilities of dogs. Right in our own backyards, or kitchens, or even beds, we have animals demonstrating the extraordinary power of evolution on behavior as well as appearance. We can even change each other's hormones, with a recent study suggesting that gazing into your dog's eyes can alter your—and their—oxytocin levels, a hormone associated with bonding and care.[39] Dogs have become really, really good at being with people.

At the same time, increased scientific scrutiny of dogs has debunked some of the myths about their behavior. For example, a large-scale study of hundreds of breeds had a difficult time finding consistent differences in personality among them, despite frequent popular characterizations of which breeds are best for what. As Lea and Osthaus make clear, we actually know relatively little about the abilities of many animals, not just dogs. Once we know more about both those abilities and about the genes and experiences that produce them, we will be closer to understanding the animals that we first domesticated.

5

Wild-Mannered:
The Other Domestics

D ogs are all well and good, the quintessential domesticated species. And we understand what both we and they get out of the bargain: wolves got the safe comfort of home, hearth, and kibble, while we humans got a companion and hunting partner. But what about all those other creatures that have come to share our lives, from camels to sheep to that internet star, the cat? Anecdotes abound, and passions can run high. Consider this comment on the blog *Why Evolution Is True*:[1] "I still fail to get my head around the concept of an 'indoor cat'. They're obligate carnivores, and instinctive hunters—far more so than dogs." Being an "instinctive hunter" is simply unchangeable, according to this view, no matter how many times a lap is sought.

But is that attitude, that being a hunter is instinctive and immutable, correct? We talk about cats and dogs as polar opposites, but what can we learn from cats, as well as all those other animals we've domesticated? How do they illustrate our ideas, right and wrong, about the way behavior evolves?

Let's return to the distinction between animals that are truly domesticated and those that are simply tame. As for the latter, the internet is full of videos of people petting tigers and draping pythons around their necks. These animals may become tame, meaning that they have been raised around humans and are far less likely to do us harm than a member of the same species snatched from the wild, but they are not domesticated. Domesticated animals have been genetically modified through selective breeding to be tolerant of people, and they are distinct from their wild ancestors. Even without being near people from birth or hatching, domestic animals are much more accepting of humans, and they associate with us in ways that wild animals do not. In his book *Why Did the Chicken Cross the World?*,[2] Andrew Lawler calls domestication "a long-term and mutual relationship, with bonds that can never fully be dissolved." Whether "never" is an overstatement or not, it is true that with a wild animal, you must start from scratch each generation.

Domestication is not, however, an all or nothing event for a species. There is no moment when a species is fully domesticated, and sometimes a domesticated species returns to a more wild state, becoming what is termed feralized, as I discuss later in this chapter in the section on chickens. The distinction can be fuzzy, and as with the dogs and wolves, our definitions might be overly reliant on our own or our culture's conceptions of our relationship with a specific animal. One person's pet may be another person's wildlife.

I am limiting my discussion to the relationship between humans and animals, though a case could be made that neither is key. Humans also domesticated plants, of course, and the progenitors of some of our crops are virtually unrecognizable as such—the ancestors of apples and bananas, for example, look like completely different species from those in the produce aisle. And some species of ants cultivate fungus in underground nests like farmers. Because the fungus is genetically altered from any form that lives in the wild, it fits the category of domesticated even though no humans are involved.

However you define it, though, domestication is not a common occurrence. Only a few dozen animals have been domesticated, and that's if you include some borderline cases like carp. Domestication of animals is thought to have begun during the Neolithic period, when agriculture emerged. Many scientists attribute the success of humanity to its ability to domesticate animals, contributing not only to our food supply but also to transportation, clothing, and spiritual practices. Farming societies, which include both domesticated plants and animals, have more surviving children than hunter-gatherer ones. That in turn led to everything from complicated political systems to the formation of cities. As a paper by Carlos Driscoll from the University of Oxford and his colleagues puts it, "Neolithic farmers were the first geneticists and domestic agriculture was the lever with which they moved the world."[3]

To see the extent to which selection has affected our barnyard and household companions, scientists use sophisticated genomic techniques to compare the genomes of populations of domestic animals with their contemporary wild ancestors, if they exist, or with DNA extracted from ancient remains, if they don't. The historical picture that is emerging is one of a back-and-forth exchange of genes between domesticated and wild animals over the centuries, rather than a single event of domestication. Some of the to-and-fro happened through interbreeding at the margins—an enterprising female pig mated with a wild boar in the neighborhood, for example. In other cases, people moved their own versions of domesticated animals as they traveled, whether on trade routes or as part of conquest, and those animals mated with the local breeds, whether domesticated or wild. Either way, it means that the origin story of many domestic animals is more complicated than first thought.

The key to all animals domesticated by humans is the ability of the animal to be around people, which means that behavior is at the heart of our understanding of domestication. At the same time, domestication illustrates how intertwined behavior is with physical attributes. Barnyard pigs both look and behave differently than wild boars. Even if, as we saw in the previous chapter, the details of a unified domestication syndrome aren't borne out by many domesticated species, it is still undeniable that being kept by humans changes the appearance of many animals in addition to changing their behavior. In many cases, the same behaviors that occur in the ancestors are found in the domesticated version, but with different ages of emergence, so that behaviors that usually disappear in young wild animals persist for much longer in their domestic descendants. Such modifications illustrate the tinkering nature of evolution—genes are repurposed, with the timing of their expression altered, rather than arising de novo in the domestic species.

Animals become domesticated in a variety of ways. They can move into our homes themselves, like cats, or even lice. Alternatively, humans can take control, either bringing animals previously hunted into captivity, as with sheep and pigs, or choosing them for other benefits, like guinea pigs for experimentation and, at least according to some people, bees for their honey. All domestication, however it comes about, is a kind of rapid-fire evolution, which means that the way behavior of domesticated animals has changed compared to their wild ancestors can tell us about how behavior writ large evolves.

Do Cats Really Walk by Themselves?

Dogs have owners, while cats have staff, or so the saying goes. Folklore about the differences between the two—and why people often have strong favorites—abound. But why are cats so much more independent than dogs, with such different behavior?

To answer that question, we need to look at the ancestry of domestic cats.[4] The ancestor of the fireside tabby or calico is the Near East wildcat, which comes from the Mediterranean, Middle East, and North Africa. It looks remarkably like its domestic descendent, though modern cats have a much larger range of fur colors and body shapes. Unlike wolves, wildcats are not particularly social, and they hunt by themselves, rather than cooperatively. This makes them somewhat unlikely candidates for domestication. Nonetheless, cats are thought to have been associating with humans for least 9,500 years, given the evidence of a cat buried next to a human in Cyprus. Other forms of art show cats in ancient Egypt, although whether these are wildcats or domestic ones can be hard to determine. Because of the skeletal similarity between the two, archaeological remains of bones are difficult to tell apart as well, and it is only with the emergence of techniques that allow examination of ancient DNA that scientists have been able to trace the genetic relationships among modern cats and their ancestors.

Scientists speculate[5] that after agriculture emerged, wildcats found easy prey in rodents near food storage, an activity that would presumably have been encouraged by farmers. The cats then became what scientists call synanthropes, which sounds vaguely insulting, but refers to animals that live around humans and benefit from them, but aren't managed or bred for particular qualities. This opportunism, as compared to the joint hunting by dogs and early humans, meant that cats kept much of their behavior, other than increasing their tolerance of people enough to get food scraps to go along with their diet of mice and rats. In turn, farmers didn't need to shape the behavior of the cats, since the cats were doing what came naturally to them.

Once cats became domesticated enough to be handled, they were moved by people along land and oceanic trade routes into the Middle East, and parts of Africa and Europe.[6] If local wildcats were in the area, the domestic versions would mate with them, so that genes from wildcats were repeatedly introduced back into the domestic animals. In Egypt, the movement of cats was thought to have been banned because of their sacred status. But those ancient cat lovers seemed to have been persistent, because genetic evidence of Egyptian cats can be seen in most of Europe. Cats were prized among Mediterranean cultures, presumably because of their rodent-control abilities. In China, cats were associated with humans from at least five thousand years ago. Their diets, as indicated by the chemical composition of ancient cat bones, seems to have had some millet in it. The millet presumably came not from the cats eating the grain itself but from the digestive tracts of the rodents that the cats hunted.[7] People didn't start breeding fancy kinds of cats until the Middle Ages, when the now-familiar tabby coat pattern emerged. Most modern breeds are still more recent. Even now, some scientists consider domestic cats to be a subspecies of the wildcat *Felis silvestris*, while others put them in their own species, *Felis catus*. In comparison, dogs are uncontested in having their own Latin designation separate from wolves.

Although domestic cats look quite similar to wildcats, it would be a mistake to think that the thousands of years of living with humans has left them genetically unchanged, in either appearance or behavior. A recent and remarkably thorough study[8] of cat genetics led by Wes Warren of Washington University in St. Louis revealed a small number of regions in the genome that differed among six breeds of domesticated cats and wildcats. About half of the genes that had diverged the most were associated with the nervous system, including memory and fearfulness, while others had to do with diet, including the reliance on meat, and the sensitivity of the senses of smell and hearing. Domestic cats seem to have higher expression of genes that are associated with reward seeking, which would be consistent with the notion that they benefited from their association with people. The authors suggest that the changes in behavior were associated with "selection for docility," but still conclude that the domestic cat is remarkably similar to its wild ancestor.

Because of their elusive behavior, few people have observed wildcats in their natural habitats, although one study[9] compared six captive wildcats at a zoo in Italy with five domestic cats in a house. The researchers watched both groups for set periods of time over several days, meticulously recording the activity of each individual.

As should come as no surprise to any cat owner, the domestic cats spent 44 percent of their time "inactive," which included both "individual" and "social" resting, the latter presumably meaning the practice of sleeping in a heap, as seen in multiple-cat households. The wildcats were inactive just 22 percent of the time, but the overall kinds of behaviors—scent marking, grooming, interacting with humans—were quite similar in the two groups. Of course, it's hard to know the degree to which captivity influenced the behavior of these wildcats; most obviously, because they were in a zoo they did not have to hunt for their food. Along similar lines, two of the house cats had been sterilized, and one of the subjects was the son of two of the others (his name is given in the paper as Standing Ovation, and I bet there's a story there), which presumably altered his behavior as well. Given these causes of variation, it is all the more remarkable that the repertoires of behavior were so similar. Cats, it seems, act like cats.

What do the shorter time since domestication and the smaller number of changed genes in cats compared with dogs mean? Do those differences between cats and dogs address the question posed at the outset of this chapter about whether cats are instinctive hunters? Can they help us understand why dogs obey, but cats take a message and may get back to you? Or, can they explain, as one author put it, "The reason why cats don't behave, while dogs do"?[10]

Probably not. Cats definitely have had less time to accommodate themselves to a human environment than dogs, which explains at least in part why fewer of their genes differ from those of their wild ancestors. Selection, whether artificial or natural, takes time to exert changes in a population, and cats just haven't had enough time. Moreover, there isn't a linear progression of domestication that all animals follow; it's not as if dogs went through the "cat phase" on their way to greater genetic change, meaning that if we just wait long enough, we will all have cats that fetch and rescue their owners from wells (okay, maybe that last part doesn't work for dogs, either). Note, too, that those genetic changes in cats are modest in physical appearance—no cat is as different from a wildcat as a Chihuahua is from a wolf—and behavior, yet again underscoring the inextricable linkages between the two. Cats may be more similar to their forebearers than dogs are to theirs, but there still is no behavioral "essence of cat" that is preserved intact regardless of how much, or how little, the physical characteristics are altered.

Cats and dogs do share one rather endearing characteristic, which is their ability to, at times, induce increased oxytocin levels in a human. Oxytocin, which is sometimes called the "love hormone," increases when people have positive interactions with others. It is part of the biology of parental care in a variety of mammals, including humans. Paul Zak, a neuroscientist at Claremont Graduate University in California, asked one hundred people to play with either a dog or cat for fifteen minutes, measuring the subjects' oxytocin levels before and after.[11] About 30 percent of the participants showed an increase in oxytocin after their session, but neither dogs nor cats were more likely to induce the response. Playing with a dog was more likely to be associated with higher oxytocin levels only if the person had had many pets over his or her lifetime.

One could make much of these differences, and studies are ongoing to see just how connected we are to either one of our most popular household animals. But cat lovers can rest assured that their pets can at least sometimes evoke the same responses as man's best friend.

Coming Home to Fight, and Then Roost

Full disclosure: after a rather skeptical start, I have come to truly love chickens, or at least their more glamorous ancestors, the junglefowl. If you think the word *glamorous* should not apply to a bird that often ends up as a sandwich in a fast food joint, read on. Chickens are having a moment, with the backyard hen making a comeback in urban as well as suburban areas. What's more, they may be showing us what happens when animals accustomed to hearth and home go back to the wild.

Chickens, it turns out, had a rather convoluted path to domestication. They have been the stuff of legends, of worship, and of kitchen ornaments galore. Yet they were domesticated not because of omelets or coq au vin, or even their willingness to live on food scraps, but because of their behavior, and their aggressive behavior at that. Acting "chicken" is an oxymoron. They have been remarkably successful animals, at least by some standards. More chickens exist in the world than any other domestic animal (presumably excluding honeybees, about which I will have more to say shortly), and they are also our most preferred source of animal protein.

The ancestor of chickens is the Red Junglefowl, a flashy and elegant bird native to large parts of Southeast Asia. Junglefowl are members of the pheasant family, and like other pheasants, male Red Junglefowl are brightly colored, with feathers that look a bit like those on a bantam rooster as well as a fleshy comb and wattles. Unlike most domestic hens, female junglefowl are drably colored, which blends in with the forest floor, and have just a tiny fringe of a comb. Junglefowl are much smaller than most chicken breeds, averaging just a couple of pounds for roosters and one to one and a half pounds for hens. Wild junglefowl are furtive animals, darting into the forest or scrub at the least provocation. They live in groups with a dominant male, one or more subordinate males, hens, and at the appropriate time of year, chicks and juvenile birds. Social status is important to both males and females; the term "pecking order" was originally coined to describe relationships among hens, because they are acutely aware of who is entitled to the better food items or resting spots and will peck at those deemed inferior.

It has been difficult to determine when chickens were first domesticated, partly because it doesn't help to find bones in human archaeological sites; the bones of junglefowl are virtually indistinguishable from those of chickens, and the two readily interbreed even today. In 2020 a group of Chinese researchers published the results[12] of an ambitious examination of the genomes from all five subspecies of Red Junglefowl along with White Leghorns (a common breed of domestic chicken) and the other kinds of Junglefowl (Green, Gray, and Ceylon). They determined that modern chickens were originally from junglefowl that lived in southwestern China, northern Thailand, and Myanmar. The current best guess is that domestication of junglefowl happened about eight thousand or nine thousand years ago, though one study suggests that bird remains from northern China that had previously been called chickens were actually pheasants, rather than chickens or junglefowl.

Like wildcats, junglefowl are an unlikely source for our placid poultry: they are skittish in the extreme, and when captured are subject to "trap death," which is exactly what it sounds like, a stress reaction to being captured that can result in them dying. But they do have one characteristic that seems to have appealed to people in several parts of the world, namely, that adult male junglefowl are quite aggressive. Wild roosters fight vigorously to gain access to hens and assert their dominance, using the long spurs on their legs as daggers. Exactly who noticed this, and who got the idea that the bird fights could be controlled and then exaggerated by humans, is unknown. Indeed, the first breeds of domestic chickens arose from roosters selected for their pugnacity rather than from hens chosen for their size, docility, or egg production. Cockfighting, not scrambled eggs, is what led to chicken domestication.

Cockfights, however, are not a matter of simply allowing roosters to do what comes naturally. In the wild, males rarely kill each other, and they use subtle indications of dominance to settle most disputes. As I mentioned in chapter 3, in the 1990s I studied Red Junglefowl behavior, keeping the birds in flocks outdoors, first at the University of New Mexico where I did my postdoctoral work, and then at the University of California, Riverside. My colleagues and I were interested in how the birds chose mates, and how male dominance was decided. We would pair males and watch as they first sussed each other up, and then walked in tandem. They next faced off with their hackles (neck feathers) raised, finally rushing at each other with their spurs extended. We stood at the ready to break up any altercations that threatened to harm either party, but we almost never needed to interfere. A winner was usually declared after one rooster chased another away, with the victor standing upright, tail held high and feathers fluffed. In New Mexico, a number of our young roosters were stolen, and we suspected that cockfighters were the thieves. It was scant consolation that while our birds superficially resembled those that have been selected for fighting, they were not the same breed, and anyone betting on our research animals would have been sorely disappointed.

Why cockfighting now not only allows but practically requires a more violent end to the battle is likewise lost in history. Modern cockfighting involves strapping metal knives to the birds in place of their spurs, which are sawn off beforehand. Victory is declared, and bets won, only after the death of a combatant. The weapons are not trivial: during a 2020 raid of a cockfight in the Philippines held in violation of COVID-19 restrictions, the blade on a seized rooster's leg apparently nicked a police officer's femoral artery, and the man bled to death.[13] Lawler[14] cites scholars suggesting that cockfighting began "as a religious practice," and the birds do feature largely in many spiritual traditions, but why a history of sacrifice would dictate a battle to the death is unclear. A cockfight is recorded from China in 517 BC, and the sport, such as it is, spread rapidly throughout much of the world and is still popular to this day. It is particularly popular in the Philippines, where it is associated with high-stakes gambling.

Eventually, junglefowl, and later chickens, were used for more than fighting. Presumably, people noticed their willingness to hang around humans and eat leftover food, and so the barnyard flock was hatched. It's worth noting that although we take the near-daily egg production of barnyard hens for granted, such reproductive behavior is extraordinarily unnatural for a bird. Most birds breed seasonally, and at the appropriate time of year lay a clutch of eggs and then incubate them until they hatch. If a predator eats some or all the eggs, many species of birds will replace them, though it usually requires some time for the female to garner enough physiological reserves to manufacture a new batch. Clutch size is a characteristic of a given species, with albatrosses, for example, laying just one egg, bluebirds four to six, and some geese up to twenty, before they start to incubate. Scientists have hotly debated the evolutionary origins of variation in clutch size, but the general principle is that birds lay as many eggs as they can eventually rear.

This is a tidy and obviously advantageous system for a given bird, but whether and how the eggs are replaced differs among species. Some birds, such as kestrels, will lay a given number of eggs and call it a day, regardless of whether a predator—or an interfering scientist— removes any of them. Others will doggedly replace egg after egg, with the ornithologists' classic example of a flicker, a type of woodpecker, which laid seventy-one eggs in seventy-two days, after which time the investigator apparently took pity on the poor bird and gave up.

Chickens, as it happens, are among the latter type, which means that as eggs are removed, the hen obligingly replaces them. This characteristic is of course essential if chickens are to be exploited for their eggs, and it means that although the fowl were originally domesticated for cockfights, they might not have been such valuable animals to keep around if they didn't also have such a useful reproductive behavior. Domestic chickens have been artificially selected to continue laying eggs to an extent that would challenge even the most determined junglefowl, of course, but it would have been a nonstarter if humans had originally domesticated, say, Mourning Doves or terns. I wonder if chickens would have been domesticated if they were only good for fighting, and not for eggs.

New breeds of chickens are still being developed, with broilers that grow ever faster, to the point where their skeletons have difficulty supporting their musculature, causing what seems to be chronic pain. Setting aside the ethical concerns about such drastic manipulations to the physiology and anatomy of chickens, it is obvious that domestication has changed many aspects of chicken biology, including their behavior.

Chickens in the Wilderness

I first went to Hawaii to look at crickets and chickens. Well, also to go to a conference, but I wanted to find a research project that could extend my stay. The crickets turned out to be a surprisingly fruitful research direction, and I am still working in Hawaii on the project that developed from that visit. The chickens were another story. As I mentioned earlier, I was working on Red Junglefowl, and I had heard that the island of Kauai was home to a population of the birds living freely in the forest that were descended from the junglefowl that the Polynesian colonists of Hawaii had brought with them well over a thousand years ago. Since the junglefowl I had been studying were captive, and since the birds are difficult to study in their native habitats in Southeast Asia, I hoped that the Hawaiian version would be a way to do some fieldwork. I also wanted to find out how the birds behaved when they weren't in enclosures at a university.

I was wrong; I couldn't do the fieldwork at all. The moment I got to Kauai's Kokee State Park, I saw poultry all right—but they didn't look like junglefowl. The hens had big combs. The roosters sported yellow legs and splotchy feathers with white spots, and even the crow sounded off; instead of the shorter notes of the junglefowl, I was hearing the cock-a-doodle-doo of barnyard fame. I gave up on my plans. You can't study junglefowl if they are really domestic chickens.

I probably shouldn't have been so hasty. It is true that the Kauai fowl are not "pure" junglefowl. But they are something that is perhaps more interesting: domestic chickens that have escaped captivity and bred with the junglefowl that were already on the island. The resulting birds are what scientists call feralized, and they represent a kind of domestication in reverse. Similar events have happened with other domestic animals, including pigs, goats, and rabbits. A handful of scientists, including Dominic Wright and Eben Gering, are studying the birds on Kauai as well as feralization in general, and they're finding that the animals challenge some of the conventional wisdom about domestication.[15]

First, in all known cases of feralization, the once-domestic animals do not just revert to a replica of their ancestral form. Instead, new combinations of genes arise, suggesting that evolution does not simply repeat itself. By examining the genome for places where DNA has remained relatively stable, the researchers determined that the Kauai chickens have genes that differ both from those in modern junglefowl and from those that have been involved in the domestication of modern poultry.

Second, the strong selection on domesticated species for just a few characteristics, like that egg-laying ability, would usually deplete the genetic variation needed for a response to new environments. While it is true that domestic animals are more homogeneous and uniform in appearance than their wild cousins, studies of feralization show that plenty of genetic variation remains. What is more, far from being helpless misfits in the wild, Kauai's chickens are doing just fine in their new habitats.

Many questions remain about how behavior in the feralized birds differs from that of either their barnyard cousins or true junglefowl. I have been particularly curious to know how comb size is evolving in the Kauai fowl, because in the junglefowl, males with larger combs were preferred by females and were also more likely to become dominant in their social group. Comb size is linked to testosterone levels, and hens can even use comb size as a signal for how healthy a rooster might be. Junglefowl hens have very small combs. In domestic poultry, hens can have very large combs, and the differences between males and females are often not so pronounced. Do the Kauai hens also favor large-combed males? Or do they like males with completely new combinations of feather colors? Maybe I have to come up with a new conference to go to as an excuse to find out.

From Ritual to Pet to Lab

How did a furry rodent go from use in religious ceremonies to a synonym for experimental subject, and a slightly derogatory synonym at that? I am talking about guinea pigs, the beloved pets of many children. It turns out that the humble guinea pig has a rather storied past, involving, among other things, tuberculosis, Caribbean islands, and George Bernard Shaw.

What we call guinea pigs are descended from South American rodents called cavies, which look almost exactly like the pet store variety except that cavies' fur is a uniform brown rather than splotchy. Guinea pigs also have slightly shorter faces and smaller brains than cavies, which is in keeping with some elements of the domestication syndrome. They do not resemble pigs very much, at least to my eye, but Linnaeus, the famous Swedish biologist responsible for developing the Latin system of taxonomy, dubbed them *porcellus*, which does mean small pig. The guinea part of the name is completely shrouded in mystery, with no one able to uncover any connection between the rodents and the African country called Guinea, or anything else that seems plausible. No hard and fast rule makes one animal a cavy and another a guinea pig, as is often the case for common (as opposed to scientific) names.

In the Andes Mountains of South America, cavies were and are used as food, which is why they were domesticated several thousand years ago, rather like sheep or cows or pigs were. Like those other domesticated mammals, domestic guinea pigs were bred to be larger and to have bigger litters than their forebearers, and they also have a wider variety of fur colors. They were also important in cultural practices, as evidenced by guinea pig remains from burial sites. A study from 2020[16] found that the animals were domesticated in at least two places, Peru and Colombia. They were transported to Europe by the Spanish conquistadors about four hundred years ago and became popular pets. England's Queen Elizabeth I was a guinea pig owner. By the 1800s, people bred guinea pigs and showed them as they did dogs and cats, and several different varieties were developed.

How did we get from show animal to lab subject? Guinea pigs became the symbol of scientific experimentation in part because they were used by the famous late nineteenth-century microbiologist Robert Koch to study diseases such as anthrax and tuberculosis.[17] They share some helpful attributes with humans, including a similar hearing system and a need for vitamin C, which make them good alternatives to other rodents. They are also docile and adapt well to living in cages. Their larger litter size meant more animals were produced for the same effort. But during the same time that Koch was active, public sentiment against vivisection—experimentation with animals—was also growing, and guinea pigs were a particular target of concern. This is where George Bernard Shaw comes in.[18]

Shaw was a staunch anti-vivisectionist and fond of the slippery slope argument, pointing out in an essay that once animal experimentation is allowed, "you will soon prove that you are justified not only in vivisecting dogs and guinea-pigs, but in dissecting every human being you can get into your power." Along those same lines, he decried "the folly which sees in the child nothing more than the vivisector sees in a guinea pig."[19] The unsuspecting rodent was thus transformed into a pejorative about science itself, and even now a reference to being a human guinea pig is not flattering.

Guinea pigs are no longer used as, well, guinea pigs very often any more. That is probably not because of the sensationalism of Shaw and his ilk, but rather that other laboratory animals, especially mice and rats, have replaced them in medical research. They still play a role in studies of allergies and respiratory diseases as well as nutritional work because of their dietary need for vitamin C. The debate about animal research in laboratory medicine continues, but, interestingly, researchers now use guinea pigs and cavies for understanding aspects of animal behavior. Because cavies and guinea pigs are so similar, one can compare their performance in behavioral tests, which provides a useful way to understand how behavior, perhaps most notably the development of adolescence, changes under artificial selection.

People probably don't think much about guinea pigs going through the equivalent of the teenage years in humans, but like many other animals, guinea pigs go through a distinct phase around the time they become sexually mature.[20] Like humans, guinea pigs (and cavies) live in social groups, and the kinds of interactions that a young guinea pig has during puberty affect what the rodent is like as an adult. For instance, although a young male guinea pig's testosterone levels increase as he matures, if he is in a group with adult breeding males, he will not behave aggressively toward them to access potential mates. This is not because he is afraid of them, or because they control his behavior, but because of a complex mix of hormonal and social signals that, in effect, say "hang on until you figure this out." His behavior then allows him to learn about life in the group, and in turn that means he can successfully be part of the colony. Guinea pigs reared with a single female during adolescence, in contrast, are aggressive toward adult males, which as you might imagine doesn't do them any good when they are wooing a mate within a group. The parallels with human socialization are interesting to contemplate.

Cavies and guinea pigs behave similarly, as you might imagine, but when and how the behaviors emerge often differs.[21] Guinea pigs explore their environment less than cavies, but while the young animals of both species play, guinea pigs are more playful as adults. (In case you were wondering, playful guinea pigs or cavies execute "frisky hops" and will suddenly run and change directions.) Many of their behaviors suggest that, as with other domesticated animals, guinea pigs retain their juvenile characteristics longer than cavies do. Both species learn cognitive tasks like associating symbols with food, and again, despite having a smaller brain size, guinea pigs show no signs of being less intelligent than their wild ancestors. Guinea pigs are also used as therapy animals, a practice that seems to be beneficial to both the people and the rodents.

Hives of Domesticity

Honeybees are many things: symbols of industriousness and cooperation, mainstays of pollination, and sources of wax and honey.

They have been associated with humans for thousands of years. But are they domestic animals? People have surprisingly strong feelings about this issue, and I can recommend that if you are ever at a loss for conversational topics, you can try asking your companions what they think. I guarantee that differences of opinion, and a lively discussion, will ensue.

It isn't just a matter of idle debate. We treat our domesticated animals differently from wild ones, and they are seen differently under the law; imagine how the courts would view a person bitten by a coyote or fox in your yard compared with that person attacked by your German shepherd. Even beekeepers differ in their views; on the website Keeping Backyard Bees (keepingbackyardbees.com), some contributors wrote that bees are not domesticated because they can return to their wild state—become feralized, as I discussed previously—very easily. Others said they must be domesticated because they have been selectively bred to be more docile, though this is not always true, as one beekeeper responded, "Bees will be bees. One day they're calm, the next day they're ornery."[22] Yet other opinions try to split the difference and call bees "semidomesticated," or "managed" but not domesticated, which likely pleases no one.

To resolve this controversy, I am happy to follow the authority of Tom Seeley, bee expert extraordinaire, who has been working with honeybees for many decades. He declared that the insects are indeed domesticated, along with sheep and goats, in a process that started at the time that agriculture began some ten thousand years ago.[23] Evidence of the long relationship between people and honeybees dates back at least seven thousand years, to Neolithic sites in Europe, the Near East, and North Africa. We tend to think of early humans as using honey to satisfy a sweet tooth. However, the beeswax from hives was perhaps just as important to them. Beeswax is easier to detect than honey in archaeological sites, because beeswax has a unique chemical composition that lingers in ancient clay vessels. This waxy signature has been found in many parts of the world, telling the story of a substance used in cosmetics, medicines, and ceremonies.

Having taken a stand of pro-domestication for the bees, I still have to admit that they cheerfully violate many of the generalizations about what domesticated animals are supposed to be like. Even after thousands of years of captivity, no domesticated breeds of bees have emerged, as has been the case for virtually all other domesticated animals. A somewhat peculiar list from a 2009 paper summarizing characteristics of species well-suited for domestication includes "males dominant over females" and "males initiate mating," neither of which hold in the matriarchy of the hive.[24] It also mentions "solicits attention" and "limited agility," the latter presumably referring to the escape ability of the animal, again attributes at which honeybees are failures.

Whether or not one concurs with this characterization, Seeley suggests that two features rescue the bees from this domestication hall of shame. The first is a proclivity for nesting in cavities like tree holes, which are easily mimicked by water jars or baskets. He imagines that enterprising swarms of wild bees took advantage of abandoned vessels, making themselves available to local people; this would also mean that honeybees in a sense domesticated themselves, or at least initiated the process.[25]

The second attribute goes back to the sometimes fuzzy boundary between domesticated and tame. It would seem that the way to a bee's heart is through her stomach. Honeybees are, in Seeley's words, "amazingly reluctant to sting once they have filled their crops"[26] (the crop is the organ where honey is stored in a bee's body). A full bee is a more placid bee, and full bees are more common under two sets of circumstances: right before they swarm (as they prepare for an energetic flight to a new home) and when they smell smoke. The latter response is one of the keys to successful beekeeping, and it may have originally benefited wild bees by enabling them to retreat deep inside a crevice or other nest location and wait out a fire with a good stash of food. How humans figured out that a puff of smoke would calm the inhabitants of a hive is a tantalizing mystery, but a 1915 treatise on beekeeping notes that "by the proper use of smoke and especially by the way the colony is handled, the beekeeper can seemingly do with his bees as he pleases."[27]

Part of why Seeley is a champion of the idea that bees are domesticated, or at least distinct from their wild progenitors, is the contrast between their lives in a manufactured hive and that in the wild. He and his students have spent many decades pursuing feral bees in the northeastern United States and studying the ways their behavior differs from the kept versions of the species. For instance, wild bees can increase their body temperature and that of the hive by shivering, which they do by vibrating their wing muscles as they gather in a tight cluster. This is remarkably effective, and it is probably how bees survive the winter in cold climates. There is a catch, however, which is that the process works best if the walls of the hive are thick, as they would be in a tree, providing good insulation. In a thin-walled commercial hive, much of the heat is lost to the outside.

Seeley advocates a kind of enlightened evolutionary approach to beekeeping as a way to promote bee welfare, in which we let bees be bees. It is not clear whether this is even possible in the modern world, where these insects have been manipulated in so many ways. To pollinate our crops, bee colonies need to be much larger than they normally are in nature, and wild bees sometimes die of disease or parasites, something that commercial beekeepers try to control and probably wouldn't view as an acceptable loss. I discuss the ways that bees medicate themselves and their colonies in chapter 11.

Bees also buck another tendency of domestication, which is a lack of genetic diversity. Through generations of inbreeding, with the offspring of just a few pairs of individuals taken for selection, animals like dogs and horses have become relatively genetically similar, which can cause genetic disorders such as hip dysplasia in some dog breeds. One might think that bees, with all the thousands of workers in a hive produced by a single queen, would be particularly subject to this problem. But a recent genetic study of bees from around the world[28] found that, contrary to this assump-

tion, the breeding of honeybees has actually increased their diversity compared to ancestral strains, perhaps because beekeepers are
prone to bring in new strains and hence allow genes in bees from
different parts of the world to mix. Commercial bees in Canada, for
example, are more diverse than the European strains from which
they are derived.

That same 1915 book on beekeeping concludes somewhat dole-
fully that the beekeeper "cannot overstep the bounds set by the
instincts of these animals."[29] This sentiment is surprisingly similar
to the one voiced about cats and their instinctive hunting abilities.
If by instinctive one means genetically determined, I would argue
that far from illustrating that it is fixed by genes, domestication
shows that behavior is extraordinarily flexible.

Does Domestication Make You Dumb?

Domestication can do many things to animals. When it comes to
behavior, domesticated versions of species differ in many ways
from their progenitors, though all are more tolerant of humans.
But scientists suggest another common thread, namely that being
associated with humans has meant a reduced evolutionary pressure on cognitive ability. There is a certain intuitive appeal to this
idea. If someone is taking care of you, you don't need to be as good
at detecting predators or finding food; hence the organ needed for
such complex behaviors will become smaller as selection favors
other characteristics, like plump breast meat in chickens or fast
running ability in greyhounds. In keeping with this notion, many
domesticated animals have smaller brains than their wild counterparts of the species.[30]

This idea is called the "regression hypothesis," although some
researchers suggest that the reduced brain size could instead be
considered an adaptation to the animals' new environment. Yet
another hypothesis is that reduced brain size could emerge from

a trade-off between investment in rapid growth and investment in braininess: the growth of one organ takes energy away from the growth of another. If individuals therefore can't do everything, and the artificial selection for characteristics that are important in domestication prioritizes breeding from bigger, or furrier, or even tamer individuals, then thinking ability simply might go by the wayside. Either way, domestication can show us how selection for behavior can go hand in hand with changes in body size and shape, or vice versa. These trade-offs have been studied most intensively in birds, particularly chickens.

Brain size as a measure of intelligence, or of anything else, really, is controversial, as I discuss in chapter 7. But differences in brain size between wild and domesticated varieties are common, and they are seen even within different breeds of chickens, with laying hens having smaller brains than other breeds such as broilers.[31] One could come up with a story about how laying hens do not need to be as smart as their meaty cousins, but, as always, the situation is more complicated. One of the parts of the brain that is reduced in domestic chickens is the cerebellum. Present in all vertebrates, the cerebellum is responsible for coordinating movement of the body. It turns out that hens housed in cages have smaller cerebellums than those housed in more open areas, suggesting that the environment has an effect on the brain, separate from any influence of artificial selection on a particular breed. Such a change need not even be passed on in successive generations, but could simply come about during the lifetime of the animal, perhaps partly depending on its genetic makeup. As always, genes and experience work together.

As I have noted before, animals can be different not because they have different genes, but because those genes are expressed— activated—at different times or to different extents. Broilers are more likely to express genes associated with being less fearful, which might indicate a relationship between selection for growing quickly, important for birds to be used as meat, and a tendency to be more of a risk-taker.[32] Such an association has been suggested for animals in general, and some support for the idea comes from studies of Red Junglefowl. Researchers in Sweden selected birds for high or low fear of humans, a bit like the Russian fox experiment.[33] After three or four generations, the less fearful birds were bigger, laid more eggs, and were more aggressive. Whether they were also better able to learn, or had bigger or smaller brains, was not investigated, though they did have higher levels of the brain chemical serotonin, a substance that is involved with fear responses and that could be a hallmark of domestication.

In another effort to get at the notion of whether being a good domestic bird is a trade-off against being smart, a group of investigators examined the relationship between learning ability and egg production.[34] Chickens were rewarded with food when they pecked at the appropriate symbol displayed on a computer monitor, and the number of tries they took to learn which image was correct was recorded. The hens came from lineages that had been selected over many generations to produce either two hundred or three hundred eggs per year, and the scientists predicted that the better layers would do worse on the test, having sacrificed, genetically speaking, brains for reproductive ability.

The scientists were wrong. Better layers turned out to also be better learners, suggesting that the presumed trade-off simply wasn't there. Regardless of whether that is because similar physiological mechanisms underlie the ability to manufacture eggs and remember which shape goes with getting a snack, or because of some other connection, the findings nonetheless underscore the complex relationship between body and brain.

That relationship is also part of why it's hard to draw conclusions about changes in behavior, or cognitive ability, based on brain size changes. Most studies use relative brain size, or brain weight as a percentage of body weight, in their comparisons. This makes sense because while the brain of a bird is a tiny fraction of the weight of that of an elephant, that comparison doesn't tell you much about the mental capacities of the animals. But under domestication, humans are selecting for disproportionately large bodies, which might mean that a supposed decrease in brain size is simply the result of the animal becoming larger through human intervention. A group of researchers in Sweden tested this idea by breeding domestic chickens with Red Junglefowl.[35] They found that domestication increased chicken body weight by about 85 percent, but brain mass only increased by about 15 percent. In other words, we've created such hefty fowl that their brains seem small, but it is a deceptive change. Domestication itself doesn't induce stupidity.

Finally, a brief return to dogs. Like many people, I started walking in my neighborhood a lot during the COVID-19 pandemic, often passing people with their dogs. One sunny morning, a woman with a small white dog nodded and smiled as she approached. Her dog immediately strained at the leash and growled menacingly at me. "He's a terrier, he can't help it," she apologized. She had a firm hold on the dog and I wasn't worried, but her comment encapsulates how we often think about the evolution of behavior in domestic animals. That four-pound dog didn't look much like a wolf, but his owner seemed to think his inner predator reigned supreme. Yet his behavior, like that of bees, guinea pigs, cats, and chickens, is both different from his ancestors and a product of his genes and his environment. Animals are domesticated in different ways for different reasons, but they all share that entanglement.

6

The Anxious Invertebrate:
Animal Mental Illness

"**O**kay," I said brightly, "who wants to pick one up?"

The students eyed me with suspicion, and then turned their gaze to the aquarium at the front of the classroom. I was the teaching assistant for animal behavior, and the lab of the day was on dominance and territoriality in crayfish. The creatures are easy to obtain from laboratory supply companies, they subsist quite well in aquaria outfitted with some rocks and a few inches of water, and they exhibit stereotyped aggressive and submissive displays that have been documented by biologists for many decades. As a way to demonstrate how to take data on behavior and analyze the results, they are perfect subjects.

Except that their lobster-like appearance extends to a pair of pincer-like claws at the front end, and the students didn't want to get pinched. I demonstrated, showing them how to grab the crayfish high on their midsection, right behind the claws so that the animals couldn't get a purchase on their skin. After a bit of practice and only a few mishaps, the students caught on, with each pair of lab partners taking a couple of crayfish to a separate tank, marking them with different colors of nail polish so they could tell the individuals apart, and painstakingly writing down movements like "approach," "claws up," and "tailflip" to determine which member could be considered dominant. It was all a pretty standard exercise, and the students ended up enthusiastically naming their crayfish and carefully selecting the right color of nail polish to suit.

What we didn't do was consider how this all went down with the crayfish. Recent work by researchers in France[1] has explored the behavior and chemicals circulating inside crayfish after they have undergone stressful events, and it suggests that the crustaceans are experiencing anxiety, physiologically akin if not identical to what we term anxiety in humans, and that it is controlled by serotonin, the same chemical used in many drugs for combating psychological ills in humans.

I will explain more about the crayfish, and just what led the researchers to call them anxious, later on in the chapter. But for now they serve to introduce the idea that if our normal behavior is deeply entwined with our biology, then we should be able to say the same about the times when our psychological state malfunctions. After all, our ideas about the source of mental dysfunction should follow from our ideas about the source of other kinds of dysfunction. What is the difference between physical illness and mental illness? And why should humans and animals share one but not the other? In this chapter I look at mental disorders in animals as illustrations of the way physical characteristics are entwined with behavioral ones, whether in sickness or in health.

We accept that our physical being has evolved, and our brain and nervous system, the source of our mental state, presumably did as well. Despite the oft-repeated "chemical imbalance" explanation for mental disorders, what do we really know about the link between the brain and our state of being? In many cases, the answer is not enough. But this question has bearing on whether we favor medicating mood, and why we think it's okay to put a Band-Aid on any wound that bleeds but think we should tough out bouts of depression as if they were the psychological equivalent of a paper cut.

An Obsession with Obsessions

Interest in whether and how mental disorders can be inherited has a long and rich history. Many early scientists, like the seventeenth-century French alienists, as those who studied the mind and its illnesses were called, were convinced that mental diseases could be passed from parents to their children. Charles Darwin became caught up in the idea; as Carl Zimmer says in *She Has Her Mother's Laugh*, "Darwin turned to humans for clues to heredity as well, but he mainly studied how they went mad."[2] A similar concern with "feeblemindedness" and its potential for being passed from one generation to another led in part to various social policies in many parts of the world, including unconscionable acts such as forced sterilization.

Darwin was intent not only on studying humans for their own sake, but also on demonstrating the unbroken links between humans and other species of animals, and thus he was fascinated by the potential for insanity in animals as well. If chimpanzees, for example, could feel emotions that were similar to jealousy or anger, why wouldn't they have defects in those responses? Without much justification, he speculated that insanity in animals, while present, would be less common than in humans. Other intellectuals at the time went further, and saw connections between criminals, animals behaving "savagely," and people who had gone mad. We certainly still talk about "mad dogs," since rabies can produce what looks like mental illness in animals. Extending this idea, it would seem odd that no other mental disorders would appear in nonhuman animals.

Current thinking about animals and mental disorders takes three forms. First is the idea that while animals can serve as models for human mental diseases, they do not suffer from such disorders themselves; they are simply convenient vessels of brain components and neurochemicals. Joshua Gordon, the director of the

National Institute of Mental Health bluntly stated, "Mice do not have schizophrenia. Or any other mental illness."[3] Some researchers even suggest that the ability to suffer from mental illness is a hallmark of being human, though this view is contested. In her book *Animal Madness*, science writer Laurel Braitman says, "Every animal with a mind has the capacity to lose hold of it from time to time."[4] Of course, this merely kicks the can down the road—which animals do we think have minds to begin with, and which do not? Pavlov, who pioneered animal psychology by studying dogs salivating at the sound of a bell, was the first scientist to suggest that animals would be useful models to study psychiatric problems in humans. Since that time, a variety of laboratory animals have been used to understand how behavior goes awry, and perhaps how it can be fixed. As you might imagine, I think it would be odd if, of all things, mental illness were the source of human exceptionalism.

The second notion acknowledges behavioral dysfunction in animals, but largely attributes it to mistreatment by people. As one interview with a Norwegian researcher of animal welfare put it, "Blame it on humans when animals become mentally ill."[5] The website Online Psychology Degree Guide (onlinepsychologydegree. info) admittedly not the most intellectually rigorous of sources, claims that "veterinarians and animal psychologists also agree that animals wouldn't suffer from mental disorders if we'd only treat them right."[6] The number of veterinarians specializing in behavior, and treating problem behavior in pets, has skyrocketed over the last couple of decades, and books about how to fix Fido's phobias have been reliable bestsellers. Although I agree that people can induce bad behavior in their pets, we can't lay the blame entirely on human error.

A third approach suggests that wild animals can indeed suffer from mental disorders, though evidence is scant for things like psychotic zebras or sloths with agoraphobia. One magazine article did include "hoarding" in hamsters as a potential disorder, despite the fact that the behavior of cramming food into the rodents' fur-lined cheek pouches is a completely functional activity that enables them to survive the winter. A 2015 BBC *Earth* story suggests, somewhat more broadly than Braitman, that "all animals with brains have the capacity to experience some form of mental illness."[7] The story also suggests that complex mental illness is a manifestation of higher-level cognition, or even greater intelligence, an idea that I will revisit later in the chapter in the context of schizophrenia. Certainly the idea that mental suffering is tied to artistry has been a recurrent one.

I am not so willing to see mental disorders, representing real suffering as they do, as a price anyone pays for heightened insight, much less for being human. My own opinion is that while we shouldn't expect to see the same disorders in animals that we do in humans, our connections with other species make the third approach the most plausible. After all, we think that other animals get physical ailments similar to those of humans—why wouldn't mental ailments be found in them as well? As I noted previously, it would seem beyond bizarre if mental illness—not tools, not language, not a capacity for empathy—were what separated us from all other animals. Following I will consider each of these approaches, but before getting to animals it makes sense to think about how we view the evolution of mental illness in humans.

The Evolution of Our Discontent

Although we no longer see mental illness as a single entity that can be passed from parent to child, like eye or hair color, we have long recognized that genes contribute to a tendency for many disorders. But two puzzles about this connection immediately present themselves. First, despite decades of concerted research, progress toward finding a single gene, or even a suite of genes, responsible for most of the major mental disorders in humans has been dismal. The same is true for our ability to isolate a particular part of the brain as responsible for a disorder, with a few exceptions such as Huntington's disease and, to an extent, Alzheimer's disease. When putatively relevant genes are identified, and hundreds if not thousands have been, each one increases the risk of developing a given disorder, such as bipolar disorder, by perhaps 1 percent or even less.

This should not be surprising. We are never going to find a single gene or group of genes responsible for mental illnesses, for the same reason we can't isolate a solely genetic cause for any trait, behavioral or physical. Genes simply don't "cause" behaviors, despite behaviors, whether healthy or not, having a genetic basis, as I explain in chapter 3. Finding a "genetic marker" for a psychiatric disorder is not synonymous with finding its genetic cause, and such a marker has almost no predictive value for identifying individuals who will develop the disorder. This does not preclude using genetic information to develop tailored treatments for psychiatric ailments, and we have made enormous strides in understanding the way that genetics underlie behavioral disorders just as they do physical diseases, but it does mean that we need to temper our expectations.

The second puzzle is an evolutionary one. If we assume that humans as well as animals evolved through natural selection, and we know that natural selection means that those individuals best suited to their environments leave more offspring, why do so many apparently defective behaviors exist? Surely it can't be any more helpful to a dog to be afraid of thunderstorms than it is for a human to fear heights. This is a form of a larger concern, namely, as psychiatrist and evolutionary medicine proponent Randy Nesse puts it, "Why did natural selection leave our bodies with traits that make us vulnerable to disease?"[8] In other words, why hasn't evolution taken care of our defects and left us perfectly adapted to our circumstances?

An early version of this question and a proposed answer appeared in the estimable medical journal *The Lancet* in 1967, when John Price from the Institute of Psychiatry in London plaintively asks, "Why then are we as a species lumbered with these most dis-

agreeable tendencies—why are we all not paragons of calm, energetic happiness?"[9] Price suggests that we can ascribe our difficulties
to a problem in our dominance relationships, inherited from our
primate ancestors, and has a rather *Mad Men* approach that does
not wear well with the ensuing decades since his paper. He claims
that "in our evolution we passed through a stage in which small
social groups were regulated by strict dominance hierarchy, much
as now exists in societies of baboons and macaques." Well, except
that as it happens, baboons and macaques are a diverse group of
primates, with a wide variety of social systems, not all of which
exhibit a dominance hierarchy at all. Further, little evidence exists
that we humans had such a hierarchy in our evolutionary history;
the family life of early humans is still hotly contested.

Undaunted, Price thinks our troubles are all rooted in problems
with our place in the hierarchy, suggesting that "in depressed patients
the irritability is manifested to 'inferiors,' such as the wife and children (or husband)."[10] Here the idea seems to be that lower-ranking
individuals can keep from sinking even lower if they are irritable in
their responses to others. Again, no evidence exists that irritability keeps a baboon from being dominated, even if said baboon is
actually in such a hierarchy. Finally, Price says, "Depression may be
commoner in females, and this accords with the fact that in animal
groups the status of a female depends not only on the status of her
males, but also on her status within her male's harem." This statement briskly ignores that human social systems show no evidence
of having been harem based. He also suggests that psychiatrists are
like the "overlords" in baboon society, which is not a term I have ever
heard used by primatologists and which does not, mercifully, seem to
have gotten any traction among people who treat mental disorders.

Although Price suggests a test of his theory, using two groups of
patients with psychiatric disorders, treating one in the usual way
and the other by reducing hierarchy, I am unaware of any efforts
ever made along those lines, which is probably all to the good. His
ideas seem at best antiquated, but I detail them because they illus-

trate how easy it is to construct plausible (at least to some) stories about the evolution of behavior without considering alternative explanations or even making sure they are based in fact. And to his credit, Price was one of the few psychologists willing to entertain the idea that animals themselves might show a form of a human disorder, even if it is a baboon sulking because it has descended from alpha to beta.

Other theories about the evolution of human mental illness have often focused on depression, perhaps because it is so pervasive, with the World Health Organization declaring it to be the leading cause of disability worldwide.[11] Those theories all posit a potential benefit to depression, or at least one that might have applied during our evolutionary history. For example, maybe depressed people can better conserve their resources in times of scarcity for use at a more propitious time, or (linked to Price's idea) perhaps being depressed means you are more likely to retreat from a fight with a more powerful opponent. Another suggestion is that depression is a kind of cry for help, so that if a bond is threatened, depression serves to keep sufferers from being socially excluded. In a related hypothesis, depressed people are less likely to take risks in social situations, something that may be useful if those situations are challenging to their welfare.

Nesse takes a different approach, one that does not suggest an advantage to the ailments themselves. He points out that saying disorders exist for a reason is vastly different than saying they exist for a purpose.[12] Illnesses and dysfunction abound, both mental and physical, and we know—or at least we should know—that natural selection doesn't produce the absolute best solution to a problem, for many reasons. A thorough discussion of these reasons is beyond the scope of this book, but a few are obvious. For one thing, bodies and minds evolve with many attributes at once. It would be ideal, for instance, to have lower backs that did not ache with age or exertion, but our spines are structured the way they are because of our bipedal stance. Such trade-offs may constrain the evolution of many things. We also get sick because many of the microorganisms that make us ill evolve more quickly than we do, and sometimes they get the upper hand.

It's also often been suggested that mental illness is one of the so-called diseases of civilization, emerging from the stresses of modern life. The notion that agriculture, or the Industrial Revolution, or the advent of other forms of technology made us worse off than our forebearers has a long history. Such notions are often accompanied by romanticizing the practices of hunter-gatherers and assuming that life closer to nature, whatever that means, is preferable. It is true that some ailments, like hypertension and diabetes, are much more common than they used to be, and that our diets of calorie-rich and nutrient-poor foods, combined with a lack of exercise, are not conducive to health, mental or physical. But we still see cancer in preindustrialized peoples, and depression, as well as other mental illnesses, occurs in cultures all around the world, even if they are not given the same term.

Instead, Nesse makes the case that "anxiety and low mood exist for the same reason as pain and nausea: because they are useful in certain situations."[13] Pain can help you yank your hand away from a fire, and nausea rids the body of harmful substances. This doesn't mean that they are pleasant, just that they have a function. More broadly, he argues, diseases themselves are not adaptive or advantageous. Instead, the things that make us vulnerable to disease, such as the aforementioned trade-offs, are the result of evolution.

Although a psychiatrist himself, Nesse takes a dim view of the way traditional psychiatry characterizes the symptoms of mental disorders. Rather than seeing such symptoms as responses produced by a particular illness, according to Nesse, traditional psychiatry confuses the symptoms with the illness itself. It would be as if we had a disorder called Fever, or Abdominal Pain, rather than recognizing that many different diseases can produce either one. He suggests that, similarly, depression is not a single disease, but a symptom of potentially several disorders. What we call mental disorders may be either extreme versions of symptoms, or system failures—dysfunction of a part of the brain or nervous system—that can manifest in several ways.

This doesn't mean that treating symptoms is not useful. Quite the opposite. Indeed, treating a symptom like anxiety without knowing the underlying cause can actually be effective. To explain why, Nesse invokes what he calls the Smoke Detector Principle. Smoke detectors are notoriously sensitive; they often sound their alarms when you burn toast. That same sensitivity means that they will be very likely to respond to even the smallest actual fire. Similarly, we have nervous systems that overreact and worry about things that actually are of little threat. This sensitivity ensures we avoid the saber-tooth tiger or dodge the approaching car, but it also allows us to overreact to nonlethal threats and do things such as spend far too much time doomscrolling on the internet. Nesse hastens to point out that even if anxiety is useful under some circumstances, that doesn't mean we shouldn't treat it. Likewise, physical pain is valuable, but there are circumstances when it's best to alleviate it, such as when we anesthetize patients during surgery.

A group of psychologists from Canada offered a slightly different but complementary approach.[14] They distinguish between mental disorders that exist despite natural selection, and those that exist because of it. The former group includes autism and some forms of schizophrenia and bipolar disorder; evolution does not produce perfection, and many characteristics in both animals and humans are present because, for instance, the genes associated with producing them are linked to genes that are very favorable. The latter group includes anxiety and depression, as well as potentially some forms of bipolar disorder. Here, the idea is that under some circumstances, exhibiting a version of the disorder can be helpful, a la Nesse's idea about anxiety and smoke detectors. Either way, these views of mental illness suggest that they would be found in animals as well as people.

A Model Mental Patient:
Too Much of a Good Thing

One of the best places to see the continuity of behaviors in humans and animals, along with their dysfunction, is in forms of obsessive-compulsive disorder, OCD. This disorder is characterized in humans by patterns of unwanted thoughts and fears, often of germs or infection or the need to order elements in the environment, that then lead to repetitive behavior like hand washing. The National Institute of Mental Health estimates the lifetime prevalence of OCD among US adults at 2.3 percent, with high costs in both productivity and functioning.[15] It is treated with a combination of drugs and behavioral therapy.

People have noticed for many years that some of the characteristics of OCD are also seen in other animals, particularly dogs. Some breeds of dogs are prone to what is called canine compulsive disorder, CCD, which makes dogs exhibit exaggerated and repeated licking, tail chasing, and other behaviors that are normal if they are only performed in limited amounts but pathological if they are performed to extremes. The name is different, because we can't know what dogs are or aren't obsessing over, but in both humans and dogs the disorder is characterized by what Elinor Karlsson, the scientist studying dog genetics I mentioned in chapter 4, calls "doing normal things too much."[16] She points out that both OCD and CCD are treated with the same drugs, and with about the same success, which unfortunately isn't very high. She and her colleagues have been studying CCD as a way to both treat our pets and to understand the genes associated with human OCD.

Karlsson and her team have identified variants of genes that affect a dog's risk of showing the disorder.[17] These genes govern the way nerve cells communicate, in particular within the regions of the brain that have to do with learning and memory. Dobermans are one of the breeds particularly prone to CCD, and so her lab initially zeroed in on testing Dobermans for the presence or absence of the genes in question. As with virtually all characteristics, however, simply knowing a dog's genetic makeup won't tell you definitively whether or not he or she will exhibit the disorder. Dogs, like humans, inherit one copy of any particular gene from their mother and one copy from their father, offering the opportunity to have both copies be the same variant or to have one of each. Of the Dobermans with two "normal" copies, 10 percent have CCD; of the ones with one copy of each type, the "normal" and the "abnormal" one, 25 percent have it; and of the dogs with two "abnormal" copies, 60 percent show CCD. Knowing the dog's genetic profile doesn't let you know the dog's behavior. This is why we can't ever expect to see a genetic diagnostic test for CCD, much less OCD.[18]

Karlsson emphasized that the connection between the gene copy and CCD is just the start to understanding how the disorder is inherited. After finding the genetic variants associated with CCD in that first group of Dobermans, you then need to get a completely new group of dogs, use their genetic information to predict which ones are likely to have the disorder, and then see whether you are correct. After that, you have to repeat the whole procedure with mixed-breed dogs and dogs from other breeds. Such a daunting set of tasks is virtually never performed with large populations of animals, but Karlsson is determined to take it on, and then some. She is using the Darwin's Dogs project, which I noted in chapter 4, in part to unravel the connection between genes and disorders such as CCD. When Karlsson is asked how many dogs she wants to enroll in the project, sometimes she says, "All of them!" with a smile. Other times she offers a more modest "Zillions!"

Dogs are not the only creatures used as models for OCD. A group of scientists from South Africa has been examining repetitive behaviors such as jumping and backward somersaulting in deer mice,[19] which are native to North America and differ from the common laboratory mouse, which is bred from the European house mouse. The deer mice can be reared in groups like lab mice, and they vary in how much of the stereotyped behaviors they exhibit. Researchers can separate them into groups that show a lot, some, or none of the actions. The mice are useful because they are easier to experiment on than dogs. Their OCD-like activities can be modified with both drugs and by changing their social environment—mice from the high compulsive group can be made to spend more time with each other than they do with a mouse that didn't show the stereotyped movements.

In people, dogs, and mice, we say that some individuals have a disorder because they do a particular behavior "too much." So how much—turning, flank chewing, somersaulting—is too much, and what does it mean to see similar dysfunction in such different animals? Nicholas Dodman, a veterinarian who has written extensively on animal behavioral disorders,[20] wonders if OCD in humans emerged from hunting and gathering, so that behaviors that were normal in moderate amounts, such as a concern with storing food, became hoarding. Since similar behavioral compulsions occur in other species, it seems unlikely that OCD emerged after the hominin line became distinct. Dodman's speculation, however, is another good illustration of how easy it is to construct a story about the evolution of a human behavior based on something that is imagined to have occurred in our prehistory. Psychologists studying both animals and humans struggle with definitions of diseases, and OCD and CCD (no one seems to call the mouse equivalent MCD) show us the blurred line not only between animal and human behavior but also between normal and abnormal behavior in both.

Even if animals exhibit symptoms that look like human mental disorders, that doesn't mean they are the same thing. Instead, the symptoms just may be alike because common evolutionary roots in our brains and bodies can give rise to different structures or behaviors. Animals often show highly stereotyped behaviors; early observers of animals were fascinated by the way that many species repeat the same behavior in exactly the same way, even if the stimulus is removed. For example, a nesting goose retrieves an egg that rolls away from her using a set series of movements with her head and neck. If a curious scientist removes the egg while the goose is midway through her actions, the goose simply carries on as if nothing had happened, carefully angling the invisible egg back into the nest the exact same way she does every other time. She does not have what I suppose we could call GCD, for goose compulsive disorder, but the stereotyped, invariant repetition of a behavior, one that suggests the roots of human OCD lie deep in our evolutionary history.

The Anxious Crayfish

It's time to return to the crayfish. As I mentioned previously, the only anxiety I observed in my animal behavior lab was in the students, but I may not have been looking hard enough. Pascal Fossat and his colleagues from the Université de Bordeaux in France say that they have demonstrated "anxiety-like behavior"—which the media quickly abbreviated to simply anxiety—in the clawed crustaceans.[21] The researchers presented the crayfish with a maze in an aquarium in which some arms were lit and others were left in darkness. Under ordinary circumstances, the animals explore the whole area but prefer the darker branches, presumably an evolved response that enables them to avoid daytime predators like herons. But if the crayfish were stressed by subjecting them to mild electric shocks, they stopped exploring the maze and were less likely to stay even briefly in the lighted arm, a response the scientists interpreted as characteristic of anxiety. That same behavior could be induced by injecting nonstressed crayfish with serotonin, a hormone also associated with human anxiety. A drug that reduces anxiety in people likewise reversed the changes in behavior when it was administered to the stressed crayfish, but the drug had no effect on unstressed crayfish.

Crayfish readily fight with each other, with a recognizable

winner and loser, and a later study by this group of scientists found that the loser of staged battles was more likely to avoid the lighted branch of the maze than the winner, a response that again could be modified with an antianxiety drug. And in 2019, the same research group published a paper[22] in which they asked, "Do arthropods [the group of animals that includes insects, spiders, and crustaceans] feel anxious during molts?"

Crayfish, along with many other members of their tribe, certainly seek out safe places when they are shedding their outer skeleton; think of it like shedding a pair of extremely tight, stiff pajamas that keep your internal organs contained, revealing a damp and flimsy pair underneath that takes time to harden. During and immediately after that transformation, the animal is extremely vulnerable to being eaten, and Fossat's group says that "molt events can be considered very stressful, which raises the question of whether arthropods fear molting."[23] They hint that the answer might be yes.

But does it, in fact, raise that question? Certainly we humans would find completely changing our skeleton unsettling, but maybe it's no different for the crayfish than it is for us changing clothes. Maybe it's like being able to put on sweat pants after struggling through a day in Spanx.

The scientists are not claiming that crayfish feel exactly what a person with social anxiety feels when walking into a room of strangers, though that is certainly the conclusion drawn in most of the media coverage of the work. The same drug affects both crayfish and humans, but that simply underscores how we arose, all of us, from common ancestors. Serotonin, sometimes called the happiness hormone, also affects feeding in *Drosophila*; urine production in assassin bugs; sleep in pigeons; social status or cooperation in lizards, reef fish, and octopus; and learning and memory in bees. In worms, it slows down movement when a hungry worm encounters food. It's all a far cry from targeting anxiety, much less happiness.

That laundry list of effects of serotonin reminds me of the old saying from Nobel laureate Francois Jacob that "evolution is a tin-

kerer, not an engineer."[24] We don't start afresh with each new species, reinventing a limb or a nerve cell as if engineering it for the
first time. Instead, like a tinkerer making something out of the bits
and pieces lying around in the garage, the same scraps of DNA in
one animal appear in a different form in a new one. This means
that we don't need to figure out whether the crayfish feel anxiety
the way people do. They almost certainly do not, given our different
environments, brains, and lifestyles—it would make me extremely
anxious to lay a batch of eggs and wander around with them pressed
to my abdomen for days until they hatch, but the crayfish seem
unperturbed. Our serotonin has simply been used for something
different from theirs.

Unhappiness at the Zoo

Having discovered that the same drugs can work in animals and
people, veterinarians have often employed them not only in pets
but also in other animals that are around people. In the book
Animal Madness, author Laurel Braitman describes behavioral dysfunction, often heartbreaking, in captive animals ranging from elephants to dolphins to pet dogs and cats. Noting that many of them
are on psychotropic drugs, she says, "Finding out that the gorillas, badgers, giraffes, belugas, or wallabies on the other side of the
glass are taking Valium, Prozac, or antipsychotics to deal with their
lives as display animals is not exactly heart-warming news for most
people who go to zoos, theme parks, and aquariums."[25] It's hard to
disagree. We like to think that animals we watch in enclosures are
going about their business, but simply made more visible to us as
observers, even though that is almost certainly not true.

At the same time, the statement reveals our ambivalence about
the distinction between mental and physical illnesses, as well as
between human and animal mental disorders. Presumably, if we
learned that the zoo animals were being treated for physical ail-

ments, if they were getting deworming medicine and having their kidney stones removed—which they are—we would wholeheartedly approve. And animals in nature certainly do suffer from a wide variety of parasites and diseases, all of which go untreated by veterinarians in the wild, and we seem fine with that as well, or at least accepting.

Why, then, do we feel so differently about treatment for mental illness, or are so convinced that behavioral dysfunction is caused by something we humans do? The answer, I suppose, is that we see the behavioral disorders as a preventable outcome of mistreatment. Many of the authors writing about mental illness in pets agree. The implication is that animals in nature never get mental diseases, which I freely confess doesn't sit well with me. For one thing, it brings us back to that separation between humans and other animals: Does it really make sense that people can have disorders like depression, which are seen in many different cultures and environments around the world, but animals never do? When in our evolution did our capacity for such dysfunction emerge? Did Neanderthals have OCD or depression? Since behavior and physical appearance and function are so closely linked, as I have been arguing throughout this book, it seems unnecessarily artificial to be okay with helping animals with broken legs but not with broken hearts. There are differences, to be sure, but they are not as clearcut as one might think.

Furthermore, if we are going to be so distressed about elephants on Prozac, how do we feel about people who need the medication? Finding out that our colleagues, neighbors, and relatives are using psychotropic drugs "to deal with their lives" on our own side of the glass shouldn't make us cringe, should it? So why would learning that animals can be helped by such drugs trouble us? Valid concerns exist about the use of drugs to treat mental problems, but why wouldn't medication be at least part of the treatment? The issue of overmedicating people to treat their psychological conditions is beyond the scope of this book, but we need to come to terms with our double standards, both with regard to human versus animal mental disorders, and with regard to physical versus mental conditions. An important distinction is that the animals cannot choose to be medicated, or made captive for that matter. This is to say the issue of the welfare of zoo animals is separate from how we feel about their mental disorders or lack thereof.

Schizophrenia

If it seems hard to say whether an animal is obsessed with a particular thought or idea, that difficulty is magnified many times over when humans try to decide whether animals suffer from hallucinations or delusions, as they would if they had schizophrenia. Scientists have known for some time that a tendency toward schizophrenia can be inherited. They also agree that many genes contribute to the likelihood of developing the disorder, and that the environment also plays a large role.

Modern genomic techniques, in which large swathes of DNA can be examined from thousands of people, are giving some clues about which genes might be important in schizophrenia, and when they might have evolved. A study led by Barbara Stranger of the University of Chicago identified genetic regions that seemed to protect against schizophrenia and found that selection favored those variants, meaning that they were more common than expected through chance.[26] Why, then, does the disease persist? No one knows for certain, but several researchers have suggested that the rapid evolution of the human brain, and particularly the regions associated with speech, may have produced vulnerability to schizophrenia. One section of the genome that evolved rapidly in the ancestors of modern humans is important in transmission of a neurochemical that regulates nerve activity, particularly speech and language, and that same neurochemical malfunctions in people with schizophrenia. Joel Dudley, a scientist involved in the work, speculated that the risk of schizophrenia might have evolved as humans became more intelligent.

All of these scenarios assume that other animals do not suffer from schizophrenia, and indeed, as I noted earlier, some psychologists say exactly that. They still, however, use animals to understand both the brain alterations that occur in people with the disease and the potential effects of antipsychotic medications. For example, a 2018 study[27] found that mice deprived of their mothers at a young age showed brain and neurochemical changes similar to those in people with schizophrenia. The mice also behaved differently compared with mice left with their mothers. This does not mean that the absence of a mother—whether you are a person or a mouse—induces psychosis. However, it does reinforce the notion that the environment and genes interact, as they do for all behaviors.

Similarly, scientists found a genetic variant in humans involving about twenty genes that changes the way the nervous system develops.[28] About 30 percent of people who show this variant will develop schizophrenia. It is possible to engineer mice with a similar genetic alteration, but the mice do not have schizophrenia themselves, and that is not why the mice were bred. Instead, these mice enable researchers to determine which drugs might be most effective at countering the nervous system changes caused by the genes, and to see how the molecular mechanisms that mice have in common with humans can help us understand the brain.

As with other mental disorders, people have tried to think of ways that schizophrenia could have been advantageous in our evolutionary history, including the idea that people with the disease became shamans or exceptionally creative individuals who fulfilled important societal roles. Even if this were true, for the disease to have been the result of natural selection, people with it would have to have more children to pass along their genes than people without it, and no evidence suggests that this is the case. Other correlations between schizophrenia and various attributes such as where people live have been found now that big genetic datasets can be screened. A genetic tendency to schizophrenia was found more often in parts of Europe that had relatively lower winter temperatures, for instance, but it's hard to even know where to start inventing an evolutionary rationale for that one. I think it is more likely that, as Nesse points out, there are limits to what natural selection can do, and eliminating schizophrenia may simply not be possible.

Schizophrenia as we characterize it in people may have emerged with the evolution of our gigantic brains, which makes it reasonable to think that other species don't suffer from it. At the same time, since we can only rely on people's reports of what they experience when they have the disease, it's hard to definitively say that animals don't have some version of similar brain dysfunctions. Part of the problem is the confusion of symptoms with disease: Is schizophrenia equivalent to the delusions and disordered speech we use to diagnose it? Lacking a definitive test that we could administer to either people or animals, we are left without an answer.

Philosophy, Veterinarians, and Alcoholism

The question of mental disorders in animals has not been of interest only to psychologists. Philosophers Krystyna Bielecka and Mira Marcinów[29] suggest that animals do indeed have what they refer to as mental misrepresentations, which means they can be just as delusional as human beings. Like other scholars, they note that depression, or behavior that looks like it, arises in animals from what is called learned helplessness. If one places animals under conditions where they cannot escape a painful stimulus that arrives unpredictably, they develop a kind of resignation in which, even after the opportunity to better their situation presents itself, they do not take it. Similarly, in humans one notion is that depression emerges partly from a sense that circumstances are so beyond one's control that effort is futile.

The philosophers argue that stereotyped behaviors in zoo animals, such as repeated pacing or swimming in circles, may well result from delusions that are hard to distinguish from what we see in humans who are likewise misrepresenting reality. They go so far as to draw an analogy to a frog that flicks its tongue out to catch a fly, expecting a tasty snack, but nabbing a bee instead.[30] This suggests to the philosophers that the frog has been delusional. I am not convinced by this, since it seems difficult to posit expectation in an amphibian, but then we may find it hard to declare for certain that another person's delusion is false as well.

On a more practical level, veterinarian Nicholas Dodman takes a kind of reverse engineering approach to mental illness in animals.[31] He has had great success in treating animals with several behavioral problems, including compulsive disorders, post-traumatic stress disorder, and a form of attention deficit disorder, using the same drugs that would treat a similar mental condition in humans. He concludes that if the same solution works in both, the cause of the problem must be the same as well. In books like *Pets on the Couch* and *The Dog Who Loved Too Much*, Dodman expounds on the idea that animals' emotions, and hence their behavioral disorders, are very much like those of humans. Although he acknowledges that "humans have the edge cognitively" (as we will see in chapters 7 and 8, that depends on one's definition of edge), he mocks scientists who, he claims, dismiss out of hand the idea that their pets or farm animals are experiencing the same emotions as humans: "At times it seems as though some scientists are the ones with conditioned reflexes, not animals. They hold stubbornly to their mistaken beliefs, all evidence to the contrary."[32]

But is all evidence really to the contrary? Just because the same drugs that change behavior in humans change analogous behavior in animals doesn't mean that we all have the exact same disease; it just means that our brains and bodies are similar, something our shared evolutionary history already told us. It seems to me that we can be similar without being exactly the same, and furthermore that we can be similar to different degrees. Not all animals are the same, and indeed a dog is different from a gazelle or a parakeet. I am not sure what Dodman would conclude about those crayfish, for example. He claims that "we possess the mental equivalent of a mainframe computer atop our shoulders, while the animals must make do with the less sophisticated but functionally similar Commodore 64 version."[33]

Some of the discussions about mental diseases in animals act as though we only have two choices: either animals are unfeeling robots that share no similarity with humans, or we are all exactly the same, with things such as jealousy in a cat or anxiety in a crayfish being indistinguishable from the same conditions in a person. I don't think that's accurate. When it comes to mental disorders in animals, and whether or how they are like the same disorders in humans, I feel about the same way I do about animal emotions, namely that they don't have to be just like the human version to exist. This is no surprise, since of course emotions are often malfunctioning in mental diseases. It doesn't make sense that humans and only humans would suffer from mental disorders. We all aren't alike physically, so why should we be alike behaviorally?

Finally, an interesting way to think about animal- and human-shared mental state comes from a 2020 examination of intoxication in animals.[34] Humans have been producing and drinking alcohol for many thousands of years, and the popular media is filled with stories of drunk moose, monkeys, elephants, and even birds and butterflies. It would be easy to conclude that a fondness for alcohol, and the subsequent dangers of its abuse, is deeply ingrained in our biology. Yet not all animals show the same predilections. Indeed, there is difficulty in using rodents as models for research on alcoholism because rats and mice will not drink enough to become intoxicated, even if the alcohol is offered in unlimited quantities.

A group of scientists from the University of Calgary decided to test an idea about how the genes that are important in the metabolism of alcohol would be expected to vary across eighty-five species of mammals depending on the animals' diet.[35] The researchers hypothesized that animals that eat fruit or nectar, which have the potential to ferment, would be more likely to show genes that would be able to break down alcohol than animals that eat other things. This thesis turned out to be only partially correct. Although many species that consume fruit do indeed have genes that help them process alcohol, those genes are also present in animals such as shrews, which live exclusively on insects and wouldn't touch a rotting fruit if they tripped over it. What this means is that alcohol metabolism, and hence an ability to become intoxicated or seek out the source of intoxication, is not one of those deep-seated characteristics that we inherited from our evolutionary ancestors. The researchers conclude that "human-level efficiencies" in alcohol digestion are unlikely in other mammals, and further caution that "it is a mistake to assume that animals share our metabolic and sensory adaptations or limitations."[36]

Such findings do not mean that we cannot compare ourselves to other species, whether in our search for models of alcoholism or OCD, but that we need to do those comparisons judiciously. As with the idea of emotions in nonhumans, our choice is not that animals are either exactly like us or completely different. We also do not need to create a club in which we admit dogs with CCD, pacing polar bears, and anxious crayfish, but exclude everyone else. When it comes to mental illness, animals can be like us in some ways and like themselves in others.

7

Dancing Cockatoos and
Thieving Gulls:
Bird Brains and the Evolution of Cognition

Has a gull ever snatched a french fry from you, or made a dive at your sandwich? Would you have been more, or less, annoyed if you found out that the bird knew exactly when you would appear, and was in effect lying in wait? Scientists in Bristol, England, recently discovered that Lesser Black-backed Gulls predictably showed up at a school just before snack time and lunch, waiting in large numbers on nearby rooftops for the opportunity to snag food from the students.[1] The birds were also able to visit a waste center at the appropriate time of day for freshly dumped garbage, capitalizing on weekends when people were scarce and hence less likely to disturb them. Both of these behaviors are strikingly different from the usual gull foraging techniques of actively searching for fish or other prey, and both only appear in urban gulls, illustrating how some creatures, at least, can thrive in human environments.

The gulls' predictive ability is impressive, but it is just the latest indication of birds behaving in a manner that can only be described as intelligent. Move over chimpanzees, dolphins, and even bonobos. Apes and cetaceans doing clever things is so mid-2000s. The new geniuses are birds, especially parrots and corvids, members of the crow family. An African Grey Parrot named Alex learned about 150 words, not merely repeating them but seeming to know

what they meant, and putting objects into color and size categories. When he died at the age of thirty-one, his obituary appeared
in the *New York Times*.[2] Even pigeons, birds that don't seem like
they would be thoughtful, can memorize over seven hundred different patterns, and can classify objects as either "human-made"
or "natural."

In case that seems too studious, consider Snowball, the dancing Sulphur-crested Cockatoo. This bird rocketed to YouTube fame
with his ability to move along with the beat of popular songs (an
article in the *Guardian* noted, "It all started, as some things must,
with the Backstreet Boys").[3] Standing on the back of an arm-
chair, Snowball produced fourteen distinct dance movements. The
authors of a paper examining his behavior note that these were
spontaneously generated by the bird, rather than copied from his
owner, "who does not make a wide range of movements when dancing with Snowball and tends only to engage in head bobbing and
hand waving"[4] (which is an accurate description of a lot of people's
dance moves, in my experience).

Psychologists think responding to music with movement is a
sophisticated form of behavior, and it is intriguing because it does
not seem to be necessary for a parrot's existence. As the 2009 paper
documenting Snowball's achievements put it, "Snowball does not
dance for food or in order to mate; instead, his dancing appears to
be a social behavior used to interact with human caregivers (his sur-
rogate flock)."[5] I am more than a little skeptical about this assessment, since Snowball doesn't exactly have a lot of opportunities to
mate whether he dances or not. Rhythmic movement in response
to sound has also been noted in chimpanzees, which sometimes
perform "rain dances" in the wild at the start of a storm.

Sulphur-crested Cockatoos have also featured in recent headlines
because of a behavior that is less charming than dancing: raiding
trash bins. The birds have lived in suburban Sydney for many years,
coexisting with humans and eating their discarded food. A study
published in 2021 by a group of researchers in Australia and at the

Max Planck Institute in Germany[6] used citizen science to establish that the cockatoos are not just scavenging but using complicated maneuvers to open the bins and get at the food inside. Flipping over the heavy lids of the bins requires a series of steps, from prying open the lid to walking around the edge of the bin. Only a minority of the birds have mastered this process. The technique varies among different neighborhoods, and the scientists concluded that the birds are learning how to raid trash from others, with location-specific idiosyncrasies developing as the cockatoos—that already have a rather bad-boy reputation for their raucous screams and flamboyant behavior—observe their companions.

What does it mean for birds to be able to do things that we used to attribute only to our closest relatives, like the apes, or to animals like dolphins that we already knew had outsized brains for their body size? How does a dancing cockatoo show us the folly of admitting only a select few into the club of intelligent species?

Something to Crow About

New Caledonian Crows look like your average crow, with glossy black plumage and a stout beak. They live in the forests of New Caledonia, a group of islands in the Pacific near New Zealand, and eat a wide variety of foods, including seeds and insects hidden inside dead wood and at the base of palms. In the early 1990s, biologist Gavin Hunt saw the crows using two kinds of tools to help them forage.[7] In addition to using twigs with hooked ends, the birds took leaves from *Pandanus* trees, which look a bit like palms, and modified them, biting off bits to create a kind of saw. Both tools were employed to fish prey out of crevices.

Although tool use by chimpanzees and even a few birds had been described before, the New Caledonian Crows take things to a different level. First, they make tools that are highly consistent with each other, like a craftsman would. Second, the different tool types are shaped in a particular way, with only the narrow end of leaf tools being inserted into crevices. And lastly, the crows use the hooks to grab onto their prey and lift it out, rather than simply to poke at it. Hunt points out that this level of sophistication wasn't seen in humans until after the lower Paleolithic era, when other aspects of material culture had already developed. Individual crows seem to learn how to manufacture tools from each other, rather than figuring out the process anew each time, something that is facilitated by them living in groups of several birds.

Since Hunt's discovery, some of the crows have been brought into captivity and studied at universities in several parts of the world. Something of a cottage industry examining their tool use and overall cognitive ability has emerged, with titles of papers about the crows' achievements containing phrases like "behaving optimistically" or using "mental representations." From a behavioral biology standpoint, they read like doting parents raving about the abilities of precocious toddlers.[8] Scientists house the birds in aviaries and present them with puzzles that require them to use increasingly complex tools, and do so in ways that would never be found in nature. For instance, one female named Betty took a wire and bent it into a hooked tool that could be lowered into a tube to retrieve a piece of pig heart (a favorite food), although wire is obviously not a part of the birds' environment and she had never been shown any before. They can even use one tool to get another tool, which in turn is employed to get food, even when the eventual use of the tool isn't apparent at the time they take it. The latter task may require the birds to mentally "picture" the outcome of their efforts, an ability that was thought by many to be restricted to humans, and is still considered to be controversial when applied to the crows.

Although they are captive, the birds seem, at least subjectively, to enjoy their tests. Betty is particularly amusing to watch in videos of the trials, as she cocks her head and industriously bends a piece of wire with her beak while holding it in her foot. One group of scientists even suggests that the birds show a "positive affective state"[9]—which so far as I can tell is a fancy way of saying "feel happy"—after they use tools. To prove their point, the researchers used wild crows, temporarily brought into captivity, that were trained to approach a box containing either a big reward or a small one, using the position of the box on the table as a cue for which one would be more rewarding. Unsurprisingly, they went to the box with the big reward faster. Then an "ambiguous" box—one in a new position that gave no clues about its contents—was offered. The idea is that birds that are more optimistic would go to the ambiguous box more quickly, while those that are pessimistic, expecting the worst, as it were, would either go more slowly or not bother checking the box out at all. So what makes a crow optimistic? If the scientists had allowed the birds to use tools to acquire food beforehand, they were more optimistic about the ambiguous box than if they hadn't been using tools. It seems that something about solving a problem with external objects primed the birds to expect a more positive outcome. Whether this applies to people being more cheerful about their future after mending a bicycle or solving a crossword puzzle is unclear.

Since the New Caledonian Crows use tools in the wild, perhaps it is not all that surprising that they can extend their skills to a more artificial situation. But other species of birds can use tools in captivity even though they never do so under natural circumstances. For example, ravens, which are members of the crow family, were given a choice of objects, only one of which could be used to retrieve food from a box.[10] The birds chose the appropriate tool even if they had to store the tool to be used another day. They could also use tokens that could be exchanged for food in a way that researchers claimed showed the birds' understanding of the future, an ability previously thought to occur only in humans and apes. Some scientists have questioned this conclusion, as with the tool-to-get-a-tool research on the New Caledonian Crows, but the birds clearly have a complex understanding of the consequences of their actions.

Similar tool use by animals that don't have tools in their environment has been investigated in Goffin's Cockatoos, small white members of the parrot family that live in the Tanimbar Islands archipelago in Indonesia. Popular as pets, the birds' abilities have also been studied by Alice M. I. Auersperg and her colleagues at the University of Vienna.[11] Like New Caledonian Crows, the cockatoos will extract food from crevices or holes in the wild, but unlike the crows, their beaks are not very suitable for holding twigs or other items that could be used to corral prey. Nonetheless, given paper or sticks, at least some individuals will modify the material and then use this tool to fish for food that is out of reach. They will even safeguard a tool by setting it aside while not using it and returning for it later. The cockatoos can use templates, paper models in different shapes, to make a tool that matches, for example, an L-shaped piece of paper they were shown earlier. The use of templates hints at the ability to form and remember visual representations of objects, something humans are known to do but other animals generally are not. And while the Goffin's Cockatoos weren't specifically examined for their joie de vivre while performing tasks, the investigators point out that the birds "partake in experiments voluntarily: they are called into the experimental chamber by name," which at least hints at enjoyment of the procedure.

The cockatoos also show us that tool use and other clever behaviors are not merely the result of contact with humans. People had wondered whether the lauded achievements of Koko the gorilla or the orcas at SeaWorld were the result of the animals' association with their human keepers, rather than glimpses into their natural abilities. So Auersperg and her colleagues compared recently captured cockatoos in their native Indonesia with birds that came from captive stock.[12] Both sets were presented with an Innovation Arena, a semicircular area that looks rather like the setup for a game show, with a different puzzle or task behind each of twenty doors. The cockatoos got twenty minutes to retrieve as many rewards as they could by doing things like turning a disc, bending a wire, or rotating a wheel.

The two groups performed equally well at the tasks, thus disproving the idea that we're the ones making birds smart, but with one caveat. It seems that it was much harder to get the wild cockatoos to do the tests in the first place. Only three of the eight even bothered, compared with ten of eleven lab-raised birds. Although the scientists discuss at length the effect of "expectancy theories on motivation . . . and persistence during task acquisition,"[13] it is tempting to be a bit anthropomorphic and simply conclude that a bird isn't interested in doing human-devised tricks if it is used to the call of the wild.

As with the cockatoos, virtually all the studies of bird cognition find, unsurprisingly, that individual birds vary greatly in their ability to perform an assigned task. Betty ran rings around Abel, another New Caledonian Crow in the same laboratory group. The differences are unsurprising because intelligence in crows or parrots, just like in people, is a result of the interaction between genes and the environment. All Goffin's Cockatoos are not equally adept at tool use, just as all people cannot solve a Rubik's Cube in the same amount of time, and for at least a broadly similar reason: their experiences, their genes, and the way that the genes develop. All act in concert to produce behavior.

So why is tool use such a focus for research, and such an apparent indicator of ability? The easy answer is that manipulation of objects used to be one of those attributes that we thought set us apart from other animals. Once chimpanzees were discovered poking twigs down holes to retrieve termites, that wall began to come down. Nowadays, we know that many animals use tools, including, as I will detail in the next chapter, invertebrates. Some of them even do so spontaneously; a captive gorilla, for instance, used a stick to get peanut butter out of a plastic dome, something that observers solemnly noted "does not resemble any behaviours observed in the wild."[14]

It seems that we find it particularly significant if animals behave like we do, and since we think humans are smart, that must mean they are too. But this seems unsatisfying to me; it's as if we are creating a club with ourselves as a president and then bringing in members that we think are qualified, based on how much like us they are. Hence the early admission of chimpanzees, bonobos, and gorillas, with later membership granted to dolphins, porpoises, and, now, crows and parrots.

Does this really tell us what intelligence is, and whether it's a quality that some animals possess and others do not? I doubt it. Behavior, remember, is not special; we could also make a club based on animals that fly and those that don't, or animals that hibernate and those that don't. Since we humans wouldn't belong to either the fliers or the hibernators, perhaps it is not surprising that we have less interest in categorizing animals according to those abilities. What the extraordinary capabilities of birds—and I do not deny that they are extraordinary—tell us is that different animals with a common ancestor millions of years in the past can evolve similar solutions to common problems. It is not surprising to see birds with sophisticated behavior, because there is no reason to think that only animals closely related to humans can exhibit it.

Featherweight Brains and Assfish

Why are parrots and crows, and perhaps the occasional gull, getting so much attention? Is it that they are smarter than other animals, or is there some other reason? Part of the answer is probably a rather boring practical point: it is far easier to have corvids or parrots in an aviary than it is to maintain a captive group of cassowaries, eagles, or even many small songbirds. The birds we see as common pets are those that happen to have young that are easy to rear without the parents (or pet owner) having to foray outdoors for an endless supply of caterpillars and small spiders. Among their other virtues, baby parrots eat seeds, not insects or worse yet regurgitated carrion, which is part of why we have stories about Polly the parrot rather than Camilla the condor.

But their ease of husbandry is only part of the story. Bird watchers have noticed the clever behavior of crows and ravens in the wild for many years, and canny corvids play a central role in the folklore of many societies. Perhaps, then, these birds have some other quality that makes them more likely to be intelligent. That idea in turn leads to an examination of the brain, both in those groups of birds and in birds more generally. We know we humans have big brains, and we are smart. Does it then follow that crows have bigger brains than other groups of birds? How does that fit in with the brains of mammals, or other vertebrates?

Scientists have known for centuries that vertebrate brains vary enormously in their size relative to that of their bodies. A 1987 paper[15] with the marvelous title "*Acanthonus armatus*, a Deep-Sea Teleost Fish with a Minute Brain and Large Ears" describes the bony-eared assfish, a creature with a brain that is less than three-tenths of a percent of its body size, thought to be the smallest brain of all vertebrates. In contrast, mice have brains that are about 5 percent of their body weight, similar to humans. Does that mean that we can safely conclude that mice are smarter than assfish (which would also make a wonderful title for a paper)? Not quite. The relationship between brain size, even as a proportion of body weight, and other attributes is complicated. For instance, shrews have quite high ratios of brain to body weight, and few have accused them of being intellectual giants.

An additional problem with assuming that larger relative brain size means greater intelligence becomes clear if you think not about comparing species, but about comparing individuals within that species. If it were always true that having a larger brain in relation to one's body meant you were smarter, it would mean that heavier people would have a lower proportion of their body size devoted to brains than slimmer people, and hence not be as intelligent, which is obviously false.

Rather than just using brain size as a proportion of body weight, therefore, scientists sometimes measure the encephalization quotient,[16] which is a way of determining relative brain size in relation to what one would predict in an animal of a given size. Imagine that you make a graph of brain size versus body size for a group of species, say mammals. Larger animals will have larger brains, and one could draw a line through the points using a mathematical formula. Some points, or species, will lie right on the line, and others will fall above or below it. The idea is that if a given species has a brain that is larger or smaller than the calculation would be expected to yield, this could provide a measure of its intelligence.

The ratio between the predicted brain size and the actual brain size is the encephalization quotient, and a large difference means that an animal has a much larger or smaller brain than one would predict from the line drawn through the points, given its body size. The quotient is not the same thing as the brain-to-body-size ratio; rather the quotient is a measure of how far a species deviates from what is expected. The encephalization quotients of mammals tend to be larger than birds or reptiles, for example. For a while scientists were busy calculating the encephalization ratios for many different animals, and found that, for instance, the quotient of a dog is higher than that of a horse, which in turn lies above that of a rat or rabbit. The measurement also yields some counterintuitive results. Gorillas, which after all are great apes like the chimpanzees, have encephalization quotients that are barely above coyotes.

Nevertheless, for a time this seemed like a way to measure something about brains that was consistent across many kinds of animals. But the quotient still turns out to be of limited use. For one thing, if you try and compare too many groups, you end up with a proverbial apples-and-oranges problem: primates and carnivores (animals like wolves and lions) both have larger relative brain sizes than cows or sheep. But if you drew a line through a group containing both the meat eaters and the vegetarians, you would end up concluding that the smaller-brained species always had lower encephalization quotients, even within monkeys or other subgroups, but they do not. What is more, finding out that different kinds of animals have higher or lower quotients really doesn't predict performance on tests of problem-solving ability or any other characteristic of intelligence except within a very limited scale.

The big problem, however, is not that aspects of the brain, whether its size or another attribute, are unrelated to an animal's behavior, or that idiosyncrasies of different species make comparisons difficult. It is that measurements like encephalization quotients or brain-to-body weight ratios are predicated on the notion that we will be able to rank species along a line of least to most intelligent, as though all creatures fall into order with humans (naturally) at the top, and the poor assfish—or worms or amoebae, depending on which living beings you want to include—at the bottom, and everything else somewhere in between. That assumption is part of what makes the behavior of the corvids and parrots seem so extraordinary; why would we expect birds, which by most accounts are "below" mammals in that manufactured ranking, to be such high achievers?

That hierarchical organization is a version of a scala naturae, or scale of nature, which I mentioned in chapter 2, and it is nowhere more apparent than in discussions of intelligence. As I said, we talk about animals being on an evolutionary ladder, with humans on the top rung and other species on successively lower ones, and that ladder can seem a lot like a school report card. A 2011 paper about measuring cognition in animals[17] noted that more "evolved animals tend to have more cerebral cortex [a part of the brain associated with complex behavior] than less evolved animals." As I already pointed out, animals don't come in flavors of less or more evolved. Those cartoons of a fish that turns into a reptile that then morphs into a bird, a mammal, and finally a human being (usually a man, and usually holding a spear) imply not just change, but advancement. That reptile, one assumes, is going places the fish never dreamed of. The implicit message is that evolution produces a new-and-improved animal as time goes on, with reptiles a more advanced notion of fishes, and mammals better still—akin to the way a 2021 model car is touted as better than the 2020 version, which is better than the 2019 version, and so on. Humans are thus the end point of evolution.

As I have already explained, the scala naturae is wrong, and those cartoons do not depict the way that evolution works. The author of the 2011 article was incorrect as well, and for the same reason. Animals are not cars, and a more recently evolved species is not an improvement on one that has not changed in millions of years. By that token, microbes and viruses, which evolve rapidly, should be the pinnacle of evolution, because they have changed into new forms literally in our lifetimes.

But evolution does not have a goal or try to improve anything. Yes, those individuals with characteristics better suited to the environment leave more copies of their genes to future generations, but everything that is alive now is just as evolved as everything else. Some animals, such as cockroaches and crocodiles, look more like their ancestors than others, but evolution has been acting on them just the same. And just as your brain does not have a tiny lizard inside, the brains of birds do not represent more primitive versions of mammal brains that were improved upon when mammals, or humans, came on the scene.

The Convergent Cortex, and Teaching an Old Bird New Tricks

Back to the question of why crows and parrots are smart, whether and how their brains differ from those of other birds, and what they tell us about the evolution of intelligence. If birds do not "rank" below mammals, then what we want to know is how evolution has produced similar capacities in groups of animals that have not had a common ancestor in at least three hundred million years. Psychologist Euan Macphail declared in the 1980s that all animal species were equally intelligent, with any perceived differences among them arising simply from what that animal happened to learn.[18] But that position is hard to maintain in the face of, for example, tool use by a cockatoo that has never seen a tool before and does not use one in its natural environment, compared with an animal that couldn't use tools if you left it in a Home Depot for a year. What other than differences in ability could account for the difference in problem-solving ability between crows and, for instance, finches?

Answering that question has been difficult in part because of a lingering belief in the scala naturae. When scientists first started thinking about the way that brains were related to intelligence, or to behavior in general, they used mammals as their guide. Mammals have very particular structures in their brains that are unlike those of other animals. The cerebrum, for example, is the part that makes a brain resemble a cauliflower, and it is where the nerve cells responsible for voluntary behavior (as opposed to involuntary behavior like breathing) reside. Mammals have a part of the cerebrum, the neocortex, which was said to be the brain's "latest and greatest achievement," according to the Avian Brain Nomenclature Consortium,[19] a group of scientists specializing in, as you might imagine, bird brains.

Thus, having a brain unlike a mammal indicated to the early researchers that an animal wasn't smart. The problem then was that the scientists also knew that birds could do things that suggested they were at least as bright as your average lab rat, and in some cases a lot brighter. How could this contradiction be reconciled? Perhaps the problem lies not with the animals, but with how their brains were described.

What one calls different parts of the brain is a really big deal, because it has implications for the way you think about the intelligence of the creature that has that brain. The mammalian myopia of early neurobiologists gave the parts of the brain in birds different names than they did in mammals, simply because they were convinced that the birds must have the Model T version of a brain and mammals the Maserati. It turns out, however, that the parts of a bird's brain are actually much more similar to mammal brains than had been thought. Sophisticated genetic and neurobiological analyses have allowed scientists to see connections between tissues and structures in different species. One such study examined single nerve fibers in three dimensions, a truly astonishing feat.

The upshot is that what had been called a pallium, a structure supposedly unique to birds, is really akin to the neocortex, and comes from the forebrain. The consortium's paper[20] is full of jawbreaker terms (piriform cortex! palaeostriatum!), but its aim is nothing short of revolutionary: by renaming all the structures in the bird brain, we can pave the way for a new understanding of brain function and cognition.

Thus armed, let's think in more detail about links between brain, behavior, and what we think of as intelligence. The social intelligence hypothesis was proposed a number of years ago to explain how evolution produced certain apparently exceptionally intelligent species, including humans. The idea is that the complex interactions that are a part of being in a group of other individuals selected for better memory and communication skills, and hence eventual intelligence. It is true that some more social species do better at problem-solving tests than more solitary ones. For instance, Pinyon Jays, which can travel in flocks of several hundred birds, outperform scrub jays, a set of related species that mostly lives in small groups. In primates, more social species do have bigger brains, or at least larger components of the brain associated with learning and memory, but the same doesn't hold for birds. Thus, the Pinyon Jays may be better at cognitive tests, but they don't have larger brains.

Many of the tests administered to animals to gauge their intelligence are themselves hard to evaluate. What does it mean to ask a crow to exchange a token for a piece of food, when they would never have to do such a thing in the wild? If an animal fails at a task in the laboratory, does that reflect more on the subject or on the test? One such study used what is called an A-not-B test, in which the subject has to figure out that the food that was in one place has been moved to another.[21] New Caledonian Crows initially didn't master the task, but it turned out that, understandably enough, that was because the birds were unused to looking at people's hands to decipher a problem. Once they were trained to look at hands, they did as well as great apes.

A more informative approach to understanding bird intelligence comes from examining not the response to a test imposed by humans but the occurrence of behavioral flexibility and innovation in the animals' natural lives. Maybe species that can deal with unpredictable environments have evolved more sophisticated brains with better cognition. But how to test that idea? Canadian biologist Louis Lefebvre hit on a clever solution: take advantage of bird watchers' obsession with reporting on peculiar behaviors in the species that they see, or, more accurately, of the willingness of scientific journals that specialize on birds to publish said reports. For instance, if a researcher saw a grackle, not ordinarily a predatory bird, kill a Barn Swallow, as P. LaPorte did in 1974, or a magpie eating potatoes, as noted G. G. Buzzard (no, I am not making this up) in 1989, the observation could be written up as a short note and published.

Using such notes from journals in North America and the United Kingdom, Lefebvre and his colleagues catalogued the species involved,[22] and confined their analysis to items where the authors used words like "opportunistic," "novel," and "unusual." Out of nearly six thousand short notes, they gleaned 322 examples of innovations having to do with feeding, whether that was eating unusual items or finding them in a novel way, as with the grackle. They then looked at the relationship between the species that were seen innovating and their forebrain size, taking into account body size and a few other characteristics. It turned out that the birds that were more likely to innovate also had larger forebrains, consistent with the idea that behavioral flexibility is linked to a measure of intelligence. An even larger study by Lefebvre and Nektaria Nicolakakis,[23] with 683 innovations published over thirty years, included nesting oddities (a kingfisher building its nest in peat cuttings in England, rather than the usual cavity along a riverbank) as well as feeding innovations. The unusual nesting behaviors were not found more often in species of birds with bigger forebrains, but, as before, the feeding innovations were, supporting the earlier claim.

The latest version of this line of research took an even broader view. If behavioral flexibility is an indicator of one's ability to adjust to a changing environment, then in the long term, maybe species more capable of coping with such changes will be more likely to survive. Simon Ducatez from the Center for Research on Ecology and Forestry Applications in Barcelona, Spain, led a group of scientists including Lefebvre to see if species that were more likely to eat unusual foods were less likely to have gone extinct.[24]

Once again, the researchers trawled the ornithological journals, this time accumulating more than 3,800 observations of over 8,000 bird species from all over the world. They matched the occasions of innovation with the classification of extinction risk for each species by the International Union for Conservation of Nature, which keeps a Red List of all threatened species. In a paper published in 2020,[25] the scientists showed that innovative species were gauged as less likely to go extinct in the wild. The innovations were many and varied, including the Rufous Treepie of India that has taken to eating candles made of wax and clarified butter that are left at temples; they will even grab the candles while the wicks are still lit. Although a willingness to eat unlikely seeming foods is not the only thing influencing extinction, the study illustrates what being smart might really mean in nature.

On the opposite side of the spectrum from extinction is speciation, the formation of new species from ancestral forms. Some groups of animals seem to do this much quicker than others. The world contains lots of different kinds of jays and sparrows, and not so many emus or ostriches, for example. All birds come from a common ancestor, but after that, some of the branches on the evolutionary tree evolved many more twigs than others. Here, too, brain size seems to play a role. Using information about the ancestry of more than 1,900 bird species, a group of researchers from Sweden, Spain, and Canada showed[26] that the large-brained lineages were likelier to have split up into many different species over time. What this means is that behavior, perhaps in the form of problem-solving ability, flexible response to the environment, or even social skills can affect how fast evolution happens.

Live Long and Wisely

If having a big brain is so advantageous, why don't all birds, or all animals, possess one? The answer is that brain tissue is expensive, requiring much of the energy an animal's body acquires to sustain it. Brain size, then, like most other characteristics, is a trade-off between the cost of maintaining the organ and the benefits it conveys to its bearers. If you are a human, the benefits of a large brain seem obvious, but why should a crow or a chimpanzee or a parrot have evolved a large brain? The cognitive buffer hypothesis[27] was developed several decades ago to explain why. According to this idea, big brains allow animals to live longer in environments that change quickly; hence the relationship between brain size and behavioral flexibility. The hypothesis means that we should also see bigger brains in longer-lived species that are in variable environments. But how can we test this idea?

Recent advances in the analysis of very large data sets containing information about many hundreds or even thousands of species, combined with information about the environment gleaned from remote-sensing technologies, has given some clues about how to solve this problem. Using satellite data from NASA on environmental variation in many parts of the globe, a group of scientists matched up brain size (suitably corrected for body size and a few other potential complications) in 1,200 bird species with whether the bird lived in a place with extreme seasonal fluctuations, or in a place where the plant cover changes a great deal during the year.[28] So, for instance, places with a lot of snow, and hence less plant cover, require their inhabitants to figure out how to get food during short days and under difficult conditions. Species with large brains were indeed more common in places where the environment is more variable, though it is not clear whether big-brained species were more likely to move to such places or whether selection caused birds already living in variable habitats to evolve larger brains.

What about the big brains and longevity part of the hypothesis? Here, too, the ability to use "big data" amassed from a variety of sources was put to use. Like other animals, different bird species vary in how long they live; parrots, for example, can live for decades, while many songbirds are lucky if they make it through three or four years. A 2020 study[29] examined 339 bird species, including both those that have young requiring a long period of care in the nest, like parrots, and those that produce chicks able to walk and feed themselves immediately after hatching, like ducks and many shorebirds. Within both groups, the species with bigger brains lived longer. What is more, parrots seem to have specific genes that are associated with longer lifespans, including those that improve immunity and help repair damaged DNA.

Interestingly, the longer-lived and bigger-brained species also tend to lay relatively larger eggs, which suggests they put more into their offspring, since larger eggs have more nutrients to give the chicks a head start in life. That fits in nicely with yet another study of bird cognition, one focusing on the crow family. As I mentioned earlier, brain tissue is expensive, and species with advanced cognitive abilities, whether crows or humans, require a long time to develop their skills, time that is generally spent in the nest (or house). Maybe, then, having an extended period of parenting facilitates being intelligent. As we have seen, corvids in general are good at solving problems, but not all crows and jays are alike. Siberian Jays, for example, are, as their name suggests, residents of a large part of the Palearctic. They live in small groups of a breeding pair, the offspring from previous years, and occasionally some unrelated individuals. The offspring can stay with their parents for up to four years, an astonishingly long time for a bird. The lingering childhood has been attributed to the high risk of predation by Northern Goshawks, which particularly target young birds. The goshawks are deterred, however, by mobbing, a group behavior that is exactly what it sounds like: several jays surround the hawk and dive-bomb it, chasing it away from their territory. Learning how to mob requires the role model of a parent, and a lot of observation. New Caledonian Crows also can spend several years with their parents, though in their case the young are busy learning how to make and use the palm leaf tools that they employ to get insects out of trees. In both cases, prolonged adolescence helps the young get a good start in life, and it is possible that in both humans and the corvids, extended parenting is key to the link between big brains, long life, and an unpredictable environment.

On, or Off, the Wall

Tool use is only one of the classic markers of animal intelligence that have been used over the years. Another one is the mirror test, mentioned in chapter 4, in which researchers place a visible mark on an animal and then show the animal its reflection in a mirror. If you do this with a chimpanzee, it will poke at the place where the mark appears, which implies that they recognize that what they see is a version of themselves. Since its development by psychologist Gordon Gallup Jr. in 1970, the test has been employed in many animals.[30] In addition to chimps, orangutans, Asian elephants, and Eurasian Magpies have passed the mirror test. Some scientists have concluded that it shows a strong relationship with encephalization quotient, and hence is an indicator of greater intelligence.

The problem is that other species you would expect to excel at this test, like some monkeys, crows, gorillas, and even (depending on the circumstances) some children, flunk it. As previously mentioned, some researchers suggest that since dogs don't use vision, a kind of mirror smell test would be more appropriate for our canine companions. It turns out that dogs can distinguish their own urine odor from that of others, and that they pay particular attention to their urine if it is altered by an unfamiliar scent—a kind of olfactory poking at the forehead spot when it's seen in a mirror. But does that mean the dogs know who they are? The discussion continues, with a 2019 paper[31] even suggesting that cleaner fish, the kind that groom the parasites off other fish species, can recognize themselves in a mirror. If that result holds up, they would be the only animal other than a mammal or bird to do so.

Eminent primatologist Frans de Waal says that it's hard to know whether the mirror test indicates self-awareness, in part because this is not an all-or-nothing capacity, with some animals crossing the gulf to awareness and others simply not reaching it. I agree; the fallacy of the scala naturae tells us that we can't put animals—or any other living thing—into simple bins where some have achieved an ability and others have not. De Waal then says,[32] "We need to start thinking more along the lines of a gradual scale." Here I would go a step further. Even a gradual scale implies that we are all arrayed along a line, where, just as the scala naturae would have it, one form leads to the next and the next. It suggests someone is at one end, and someone else is at the other. But evolution doesn't form a line, it creates a bush, with messy, branching stems and long distances between groups that then evolve similar properties. The animals, including us, do similar things because selection has acted on us in similar ways, not because of our ancient mutual ancestor. Hence the extraordinary capacities of corvids, parrots, and many other animals that are not closely related to humans.

If You're So Smart . . .

Of course, behavioral flexibility and toolmaking are not automatically the same thing as intelligence. Humans are pretty flexible, and we use tools, so we tend to look favorably on other animals that do the same. But defining animals that are similar to us as intelligent simply because we think we're intelligent is circular, to say the least. And as we will see in the next chapter, even insects, which have often been dismissed as robotic automatons incapable of true intelligence, can show remarkable feats of learning. What the clever actions of the birds illustrate is that we do not have just two choices: animals that are just like us, or animals that are simple machines incapable of any intelligence at all. Instead, birds have evolved similar behaviors through alternative pathways. Starting as dinosaurs—and birds really are dinosaurs, after all—they met challenges in their environments with flexibility, just as some other animals, including some primates, did.

This realization also gets us away from the game-show contestant view of animal intelligence, in which we try to rate animals for how smart they are. We don't try to rank animals based on their kidney function (though, seriously, kangaroo rats can exist without ever drinking water, which is a fairly amazing feat). Why should we try to do so on the basis of intelligence? Yet the desire never seems to go away. A paper[33] published in 2019 by an astrobiologist from the Technical University Berlin posited that we should look to naked mole rats as harboring latent capacity for getting on the list of smart animals.

Naked mole rats are rodents, as you might expect, and whatever you think of their intelligence, they would never win any contests for beauty. Hairless, as the name suggests, with elongated front teeth they use for digging, mole rats live in colonies underground, feeding on roots and tubers and defending their groups against the burrowing snakes that are their main predators. They have been the subject of much study because they have social systems akin to those of bees and wasps, with a queen that produces all the offspring and workers that take care of them. The author of the paper seems quite taken with the creatures, suggesting that their "active lifestyle" and "hygienic behavior" are "feature[s] of intelligent organisms,"[34] noting approvingly that they defecate and urinate in designated areas. He does note regretfully that mole rats have poor dexterity, and that they lack the ability to use fire, which on further reflection seems not a bug but a feature, given how difficult it would be to control flames in underground tunnels. The paper concludes by wondering if the naked mole rat "may progress toward truly advanced intelligence."

While naked mole rats are an oddball candidate for the next intelligent creature to follow crows or parrots, an enthusiasm for them points to the futility of trying to array animals as if the world was a giant SAT test. Mole rats are, indeed, good at being mole rats, but that doesn't mean they win some global contest. The crowdsourced question and answer site Quora recently featured the query, "Why is there only one intelligent species? Why aren't other species intelligent?"[35] The answers range from "Human brains would be wasted in other animals" (which begs the question of whether they are wasted on us, too, and who is doing the wasting) to "If you interpret the definition of 'intelligence' broadly enough, you can assign it to a jellyfish too" to "Mice are intelligent at mouse stuff" to "We got lucky." This last is debatable, but certainly the answers point to the realization that animal intelligence may be more complicated than is sometimes realized.

And speaking of being intelligent at mouse stuff, it's worth noting that there may be hope for rodents, which have never been on anyone's list of Most Intelligent. One of my colleagues at the University of Minnesota, Emilie Snell-Rood, led a study[36] comparing the size of brains in mice caught in urban environments with their counterparts of the same species collected in rural areas. The results attracted the attention of the National Public Radio show *Wait Wait . . . Don't Tell Me!*, in which listeners can call in to participate in a limerick fill-in-the-blank contest based on the week's news. In August of 2013, one of the limericks was as follows:

Not just must we run from the kitty
but we must seem urbane and quite witty.
A quaint rustic charm might do well on a farm
but we mice smarten up in the . . .

Listener Stephanie Garber from Florida correctly guessed that the missing word was "city." The host of the show, Peter Sagal, accurately summed up the study, saying, "Biologists at the University of Minnesota have found that city mice have bigger brains than their country cousins. That's because they say city mice have more to adapt to. In the country, it's Tom and Jerry, in the city it's Tom, Jerry, and Crazy Hobo with a Gun." This is not exactly how Emilie and her coauthor Naomi Wick put it, but it is reasonably accurate. The mice collected in urban areas indeed had bigger brains, which the scientists suggest arose because the greater levels of variability in cities selected for behavioral flexibility. Mice that could cope with more novelty in their environments did better and had more offspring than their counterparts that were in a more predictable habitat. Sagal went on to say, "The worst part is that city mice are also elitists, out of touch with real American mice. They only eat trash from Whole Foods." I have no comment on that.

8

A Soft Spot for Hard Creatures:
Invertebrate Intelligence

I t is a tale with all of the ingredients necessary for a hero's saga: the little guys under siege from a ruthless, violent foe; the clever ruse that saves the society from attack; and the realization that this had been going on for millennia in a little-known part of the world. True, the "little guys" are bees, but that fits in with a recent trendiness of backyard beekeeping and love for pollinators. A little harder to take, perhaps, is that the defense of the colony involves balls of dung.

The bees in this story are close relatives of the European honeybee (*Apis mellifera*) most familiar to us; they are still considered honeybees, but they are *Apis cerana*, sometimes called the Asian or eastern honeybee, and they are smaller, with more pronounced striping on the abdomen. They live in the same kind of hives as their European cousins, however, and in parts of Asia, including Vietnam, where this story unfolds, they are under threat from *Vespa* hornets, a kind of wasp many times larger than the bees. (Although they are closely related to the "murder hornets" making headlines, this is not the same species.) A 2020 paper[1] reports on the interaction, with the description of the hornets' strategy making it clear where the authors' sympathies lie: "During this slaughter phase, each hornet can kill thousands of bees, and, collectively, a group of hornets can obliterate a colony's defensive force within a few hours. When resistance ends, the occupation phase begins: hornets enter the nest, begin guarding it as their own, and shuttle brood back to their own nest to feed their young."

It had been known that the bees could defend themselves by, for example, surrounding a wasp and suffocating it in a ball of bees, but that wasn't going to work with this hornet attack. The recent study showed that the bees use a more novel approach. Scientists had seen small dark spots at the honeybee hives' entrances, which local beekeepers said were water buffalo dung. The bees gather the dung at water sources, and carefully place it around the hive; if enough spots are present, the hornets are deterred from attacking, for reasons that aren't clear. The authors of the study call the fecal material a tool, recognizing that this designation is a bit of a reach, but saying that "the fact that the bees are collecting something from the environment, holding it, manipulating it and changing the character of the thing that they're applying it to, makes it a tool by virtually every definition."[2]

If the bees aren't convincing, or if "pellets of animal poo," as the *Guardian* called them,[3] are a bit too unsavory, another example of recent insect tool use comes from ants that make siphons out of sand. The particular ants under study are black fire ants, which like many of their kind will drink sugar water. Most of the time, ants can simply float on the surface without risk, because their waxy coating keeps them from sinking. But when some diabolical scientists from China and the United States added a surfactant—a soapy material—to the water, the surface tension of the water dropped and the ants began to drown.[4] Undeterred, the ants began to plaster sand grains supplied by the scientists along the sides of the container with the sugar water, which allowed the solution to wick up the sides so they could safely drink. As the surface tension got even lower, with more ants at risk, the insects constructed more elaborate sand structures, enabling them not only to avoid drowning but also to get access to more of the tasty liquid.

The sand, too, was deemed a tool, and much was made of the flexibility of its use, since the ants could change what kind of siphon they made depending on the perceived risk of the situation. The authors of the study claim that the behavior indicates "high cognition," which proved to be a sticking point for some. Nevertheless, at least some readers commenting on a summary of the work that appeared on the website Why Evolution Is True[5] not only bought that the ants had advanced thinking ability but also leapt from there to the notion that the ants, or perhaps their colony, were conscious. According to one, "Such a subset of the colony implementing higher consciousness would look like a cluster of ants interacting furiously with each other, to no obviously visible purpose."

Of course, that is exactly what ants look like to me much of the time, so it isn't clear how that gets us to consciousness, or the lack of it. Still, the ants, as well as the bees and a number of other examples of invertebrate intelligence I will bring up in this chapter, point to a conundrum in how we think about the evolution of behavior, particularly behavior that suggests that an animal is aware of what it is doing. We marvel over crows using tools or chimps exhibiting empathy, yet we are still taken aback by invertebrates doing similar things, at least if we think that only some animals—and humans—have a special ability to perceive and interact with the world.

But because behavior evolves the same way that physical characteristics do, our surprise may be uncalled for. Complex behavior in invertebrates arises the same way that complex physical structures do, or complex physiology. Selection can produce flexible behavior, that hallmark of intelligence, just as it can produce digestive systems that can accommodate a wide variety of foods. The gut doesn't have to think about how an ice cream cone is different from a steak to render each into its components.

Invertebrates, then, are some of the best subjects for studies of how behavior evolves. In addition to my simply really liking them, as the chapter title suggests, I am enamored of invertebrates as a

way to understand behavior, because we can actually induce evolution in them without having to wait for centuries or millennia to
pass as we would if we were examining generations of most mammals or birds. Their small size and short lifespans challenge the
conventional wisdom about which animals do complicated things.
The sheer number of different kinds of invertebrates, from shellfish
to worms to my favorites, the insects, provide a gigantic palette for
nature to show its variety. As the philosopher Peter Godfrey-Smith
said[6] in his book *Metazoa: Animal Life and the Birth of the Mind*,
"Arthropod evolution has been exuberant for half a billion years."
And they ultimately make us rethink who we put in that special
class of intelligent creatures, and why.

Socially Inept Bees, and Crabs in a Maze

Honeybees are often considered the very model of cooperation.
Along with ants and wasps, they are called the social insects, after
all. They are also often dismissed as all alike, tiny cogs in a big exo-
skeleton machine. But it turns out that not all worker bees are
equally responsive to their hive mates, and an unlikely link between
humans and bees is responsible for the variation.

Ordinarily, bees from one colony will attack a worker bee from
a different colony, sometimes injuring her if she is introduced to
their hive. In contrast, they will fawn over a young queen, even if
she is a foreigner, and try to feed her. Researchers at the University of Illinois at Urbana-Champaign tested bees from seven different colonies by introducing them to either the larval queen or the
strange worker.[7] Some bees were more enthused, so to speak, about
attacking the apparent intruder than they were about nurturing
the potential queen; with others, it was the other way around. Most
of the bees reacted in some way to at least one of the two situations,
but about 14 percent didn't respond to either one, sitting around
on the sidelines while their nestmates rushed around acting like

guards or nurses.The scientists then dissected the brains and nervous systems of the bees, and used sophisticated genetic analysis to see which genes were most active in the nonresponsive bees. More than a thousand genes were regulated differently among the unresponsive bees, the nurse bees, and the guards. The subset of genes that was active in the nonresponsive group of bees turned out to be similar to sets of human genes implicated in autism spectrum disorder, but not in schizophrenia or depression.[8]

Now, no one is concluding that bees are autistic, or that autism in humans is caused by the exact same process that produced bees indifferent to the plight of their coworkers (though researchers are excited about the possibility of using honeybees to test ideas about autism and the nervous system). Gene Robinson, the director of the lab in which the research was performed, said in an interview,[9] "We do not want to give the impression that bees are little people or humans are big bees." It's a point that seems like it should not have to be made, but does. Instead, the research points to the seamlessness of behavior in many living things. Like the mantids that I discussed in chapter 2, which can fine-tune their behavior according to the task at hand, the bees are neither automatons nor exact replicas of humans. What is more, even creatures we deem to be identical robots can show remarkably subtle differences in behavior.

Bees are favorite study subjects for learning in insects, perhaps because they are important to humans and perhaps because they are simply interesting and seem to be able to do things that are much more complex than one would imagine a creature with such a tiny brain could manage. They can, for example, recognize individual human faces, and they can discriminate among many different flowers when searching for food. The latest bee achievement has to do with a task many humans find challenging: doing mathematics.[10] Let me hasten to point out that bees are not sitting in their hives solving differential equations, or calculating the volume of honey that will yield the most offspring for the following year. But they do seem to be able to associate symbols with numbers, at least with the numbers two and three. Such a numerical sense had been demonstrated in African Grey Parrots, chimpanzees, and rhesus monkeys, but never before in an invertebrate.

The bees in the study were trained to match a symbol, say a T-shape, with two or three objects, or, vice versa, to associate the number of objects with a symbol. It took around fifty trials for the bees to figure out the task, but after that effort, they could correctly match the two 80 to 90 percent of the time. Trying to teach the bees to reverse their task, so that if they had started with the symbol, they were then supposed to match the number, proved less success-ful. The scientists who trained the bees speculated that "an approx-imate number system"[11] exists in both humans and bees, which presumably means that many other species have similar abilities.

Bees are not, however, the only invertebrates that can learn to perform complex tasks. Crustaceans, the animal group that includes crabs and lobsters, are not exactly stars in the brain and nervous system department. For the purposes of comparison, a crayfish has just ninety thousand neurons, whereas a honeybee has over a million. But as we've seen before, brains aren't everything, and animals are often good at different tasks; no one would ask a bee to bend a wire to get food out of a tube, and crows would prob-ably never learn how to gauge the amount of nectar in a flower. A group of researchers in Swansea, England, took the ordinary Euro-pean shore crab and gave it a classic learning task, namely, a maze.[12] Rats, of course, are practically synonymous with mazes. They have lived among humans for millennia, and we know that rats are good at navigating complicated passageways. But crabs also live in envi-ronments with lots of twists and turns among rock crevices, yet their abilities hadn't been tested. The crabs got a rather advanced version of the maze test, with a crushed mussel as a food reward at the end. Virtually all of them figured out how to get to the food within twenty-five minutes, taking several trials to do so with-out errors. After six weeks, during which time the crabs were not tested, they could still find the endpoint of the maze within eight minutes.

Fruit flies are even less likely candidates for flexible learning, having been the poster animals for genetically based behavior for over a century. As I mentioned in chapter 3, they were one of the earliest subjects for the examination of the genetic basis of courtship, performed by Margaret Bastock. Perhaps because of this iconic history, no one spent much time considering whether their behavior was anything but strictly genetically determined, meaning there could be little opportunity to alter what they did in the face of changing circumstances.

Yet even the fruit flies show us the inescapable interaction between genes and the environment. Male fruit flies will court females without any prior experience if they have a particular gene called "fruitless." If male fruit flies that lack the gene are reared by themselves and then presented with female fruit flies, they stand there, apparently dumbstruck, like kids at a middle school dance. But a 2014 paper[13] found that the situation is more complicated than that. If the males without the fruitless gene are raised with other fruit flies, they can learn to court females, through mechanisms that are not yet understood (though one wonders what the equivalent of adolescent sharing of sexual misinformation might be for fruit flies). The point is that even a behavior so basic and essential that it seems as if it must be completely instinctive, especially in a fly, can be modified by experience.

What Puts the Us in Octopus?

Tell me, O Octopus, I begs,
Is those things arms, or is they legs?
I marvel at thee, Octopus
If I were thou, I'd call me Us.[14]

When he wrote this poem, Ogden Nash presumably meant that an octopus's collection of limbs made it seem like multiple animals. But one could also call the animals "us" for other reasons. If the animals-that-are-practically-honorary-humans bench has had to make room for first bonobos and chimpanzees, and then crows and parrots, the latest reason to move over is a sticky one, literally. The last few years have been full of stories about the wonders of the octopus, with its Houdini-like ability to escape confinement, its apparent awareness of people, its cleverness at solving puzzles, and, as the most recent news relates, its irascibility in occasionally whacking fish for no apparent reason.

In his book *Other Minds*,[15] philosopher Peter Godfrey-Smith beautifully contemplates how octopuses are and aren't like us, and how their complex behavior suggests a connection to us that is absent in most other invertebrates. He suggests that losing the hard outer coverings of snails and other mollusk cousins enabled the octopus to evolve a more complex nervous system. Their "different embodiment," with a shape that can grow, change color, shrink and reform in a moment, allows octopus to exist "outside the usual body/brain divide."

Does that squishiness of form mean that octopuses are smart? Maybe. As Godfrey-Smith says, "When tested in the lab, octopuses have done fairly well, without showing themselves to be Einsteins."[16] They can learn to operate levers to get food, but so can many other animals, including that workhorse of psychology, the pigeon, which no one has ever proclaimed to be a genius. Some researchers suggest[17] that a few of the octopuses' activities fulfill the requirement for play in animals, another one of those characteristics we like to believe is restricted to humans and just a few select creatures like monkeys or dolphins. A slight snag is that play in other animals is usually different in juveniles and adults, while in the octopus no such demarcation exists; the authors of a paper on play explain this away by saying that "the standard concepts of juvenile and adult life found in vertebrates do not easily translate to cephalopods."[18] Perhaps. But what else doesn't translate as well?

It is worth thinking about what octopuses can and can't do because it forces us to think about what we really mean by intelligence and what else has to accompany it, whether that is language (presumably not, since octopus do not have complex communication systems), intricate social interactions (ditto), or something else. We make a lot of assumptions about how much animals have to be like humans to be considered intelligent, and octopuses give life to those assumptions with every rippling shape-shift. As Godfrey-Smith also says, they "have an extraordinary sensorium and an anarchic bodily embrace of novelty, but they are not, for the most part, ruminative and 'clever' sorts of animals."[19]

One of the ways that octopuses defy our ideas about intelligence is that they, along with most of their cephalopod cousins like squid or cuttlefish, live very brief lives, perhaps four or five years, often much shorter. My friend Jody was telling me about watching *My Octopus Teacher*, an award-winning 2020 film about a man who spends a year watching and interacting with a female common octopus as he dives off the coast of South Africa. At the end of that period, the octopus reproduces and dies. "I was crying, Tom [her husband] was crying . . . how can they just have their babies and die?"

I shrugged, not because octopuses do not move me, but because, as I told her, dying after reproduction is pretty common in animals, especially invertebrates. Cockroaches, I helpfully pointed out, do much the same thing. Jody glared at me. "Do they really? Well, thank goodness for that!"

Is it that we think an animal that seems to enjoy life should be able to do so for longer? Or is it just that since our own lives are longer, we want the same for an animal with which we sympathize? You could just as easily ask why mayflies get a lousy twenty-four hours while tortoises can persist for centuries. Godfrey-Smith says that he assumed cuttlefish and other octopus relatives were old, "partly because they *seemed* old; they had a worldly look."

Is the question why they die so young, or why they ever became smart? A 2018 paper[20] has the rather plaintive title "Grow Smart and Die Young: Why Did Cephalopods Evolve Intelligence?" The authors, led by Piero Amodio of the University of Cambridge in the UK, note that if one relies on primates to think about intelligence, one gets a tidy story about how animals become smart in one of two ways. First, a complex environment may require flexible foraging strategies, such as group hunting. Second, as I have already mentioned, living in a group makes it necessary to communicate with the others and remember who is on your side and who is not. Either option you choose paves the way for a story of big brain evolution and culminates with human societies as we now have them.

But as the paper's authors point out,[21] many other vertebrates, especially corvids, as we saw in the last chapter, also seem intelligent, even though they have much different brains than primates. Maybe, then, the same selective pressures were at play, with the common thread being a long life with an extended juvenile period and a lot of parental care to allow the youngsters to learn the ropes of both finding food and navigating social politics. A big brain along the lines of the primates may not be required. Except—here again, the octopus and other cephalopods defy the norm, since their eggs float away unattended and mothers die. The authors then suggest that perhaps the octopus's sexual cannibalism, in which females sometimes eat their mates, has selected for fast decision-making and flexible mating tactics, but this seems like a last-ditch effort to me. After all, their non-robotic nature notwithstanding, no one is rushing to champion praying mantises, the queens of sexual cannibalism, as geniuses either. It also feels, shall we say, troubling if we've come to the point of suggesting that eating your mate is what makes a species smart. Other researchers would remove the "social complexity" part of the equation for evolving intelligence, noting that in humans, IQ and social ability are not correlated, but this, too, seems like special pleading.

Anyway, it turns out that all is not wonder and grace in the octopus world. In December of 2020, at the end of what was arguably one of the roughest years in recent history, scientists reported[22] on octopuses using one of their arms to punch—their word—fish. Some of the blows seemed to be associated with the octopus shoving the fish out of the way so that it could obtain its prey, but a few other incidents—two out of the eight that were recorded—were more inexplicable. A possible explanation of the latter, according to the authors of the paper, was "spiteful behavior, used to impose a cost on the fish regardless of self-cost." Other options were mentioned, but the spite idea immediately captured media attention, with the *New York Post* headline "Octopuses Spite-Punch Fish, Who 'Don't Like It,' Study Finds" and even the *New York Times* offering videos titled "Eight-Armed Underwater Bullies: Watch Octopuses Punch Fish." Comments from readers of the *Times* ranged from congratulatory: "If you were as intelligent as an octopus, and these annoying and stupid fish just wouldn't do what you want, you might resort to smacking a few of them, too" to amused: "Octopuses are highly sensitive animals and as such are easily offended" to critical: "Calling this bullying or lashing out is premature, anthropomorphic and stupid."

So, are octopuses us, or not? To me, octopuses illustrate the perils of reverse engineering an explanation in evolution. We assumed that if humans are smart, and that we have certain qualities like longer lives, bigger forebrains, or complex social lives, then those qualities are necessary for our intelligence. But we then run into trouble if the creatures we think are intelligent lack such qualities. Instead, maybe we can acknowledge that intelligence isn't measured the same way in all creatures—something that sounds facile, but I don't think is. This doesn't mean that all animals have the same abilities, a la Euan Macphail, the comparative psychologist I mentioned in the previous chapter, but that those differing abilities evolved, just like physical characteristics, in the environment where a given species exists.

The Really, Really Itsy-Bitsy Spider

Octopuses have eight limbs, and we think they are sinuous and graceful. Spiders have eight limbs and we think they are horrid and creepy. This seems profoundly unfair. It is particularly unfair because in addition to being just as agile in their movements as the lauded octopuses, spiders show us the limitations of the nervous system, and just how the body limits the mind, or at least the way we behave.

Bill Eberhard is a biologist with the Smithsonian Tropical Research Institute and the University of Costa Rica, and he has always been ready to question invertebrate conventional wisdom. He tends to study animals that other people dismiss: earwigs, wasps that parasitize other insects, and, yes, spiders. His work on the latter has included spiders' kinky sexual practices (he describes the sound that one species makes as "resembling squeaking leather," which in an earlier book I said was like spider porn, if such a thing can be said to exist).[23] He has also asked a question to which spiders are perfectly suited: Does being small mean they are dumb?

Many people have tried to make connections between brain size and behavior, especially intelligence, as I discussed in the previous chapter, but few have drawn this comparison out to its logical conclusion: Are there animals that are so tiny that they are almost, as the saying goes, too stupid to live, or at least to do complicated tasks, or tasks requiring flexible responses? Even Darwin wondered about this possibility, especially when faced with the complex behavior of ants; in his 1871 book *The Descent of Man, and Selection in Relation to Sex*, he mused, "The brain of an ant is one of the most marvelous atoms of matter in the world, perhaps more so than the brain of man."[24]

How might such small-scale intelligence be possible? Do minuscule animals pay a price for their lack of brain capacity? Among vertebrates, smaller individuals tend to have larger brains relative to their body size, a generalization called Haller's Rule, after Albrecht von Haller, an eighteenth-century Swiss physiologist. The rule seems to work for many invertebrates as well, but because neural tissue is expensive for an animal to maintain, it wasn't clear if there is a lower limit to the evolution of brains capable of driving complex behavior. Very small animals may well have different constraints on the way their brains can manage their behavior, but few scientists have tried to apply Haller's Rule to creatures without backbones.

Eberhard has studied the spider *Anapisona simoni*, the adults of which weigh less than a milligram. To put this in perspective, that is lighter than a single staple, or an inch of sewing thread. Yet inside their compact form is enough nervous tissue to enable the spiders to produce orb webs, the silky wheel that entraps even tinier prey. Eberhard wondered if the extremely difficult process of weaving a web was more of a challenge to the minuscule *Anapisona* than to three other spiders that weighed anywhere from ten to ten thousand times their size (which still does not give them horror movie status—a single milligram is that tiny). He meticulously compared the species' ability to adjust the space between loops, to construct different angles between the spokes of the web, and to place the sticky lines of the web, the ones used to snare the prey, at exactly the right spot.

It turned out[25] that although the miniature species had a bit more trouble adjusting its web to different conditions, by and large the small spiders were as capable as the larger ones. How do they manage that? Possibilities include cramming brain tissue into places where it is not usually found; some of the small spiders have brain tissue that spills over into their legs, giving, as Eberhard and his colleague Bill Wcislo state,[26] "new meaning to the phrase 'thinking on your feet.'" It's a mystery how this leaves enough room for other important organs, like those for digestion. It also poses questions about just how little tissue is required to run an animal at all, since the laws of physics limit just how small neurons can become. And if that isn't enough, Eberhard also reminds us to think of the children, so to speak; those teeny tiny spiders have even teenier babies, and they, too, have to catch their prey.

This work brings up the question of how brains and behavior are connected in insects, the way they are in birds, as I discussed in the previous chapter. A comparison of ninety-three species of bees from North America and Europe[27] found that body size was the best predictor of their brain size, as it is in many other animals. But bees vary in many aspects of their biology: they live in different places, feed on different things, and may be solitary or live in large colonies. It turned out that specialized species, such as those that need to eat only one kind of plant, had large brains relative to the size of their bodies, as did species that only had one generation per year, rather than having a queen with a lifespan that overlaps with her daughters and granddaughters. The more social species did not tend to have bigger brains, a surprising departure from what we know about humans and other primates, where the hypothesis that the pressing need for social intelligence led to our oversized brains. Insects, once again, break with convention.

Some people have pointed out that if we want to think of the brain as a computer, bigger should not be assumed to be better, since of course modern computers are dramatically smaller than their twentieth-century forebearers. Newer models are more efficient, and with more sophisticated technology, making a large number of components unnecessary. At the same time, Lars Chittka and Jeremy Niven, biologists from Queen Mary University of London and the University of Cambridge, respectively, argue[28] that this analogy is flawed, because nervous systems and brains, unlike computers, are not designed for a given purpose, but evolve through the "tinkering" process I noted earlier, with selection never starting from scratch but always using the components already present. Some parts of our neural anatomy, they point out, "have been retained since the Cambrian. Arguably, all extant [existing in species alive today] nervous systems are success stories; no single one is inherently better than any other."

This evolutionary history, along with the different environments of different animals, should make us wary of generalizing about what we mean by intelligence. Chittka and Niven note[29] that insects and other invertebrates can reveal the circularity of reasoning about brains and being smart; one psychologist decided that learning speed is a poor indicator of intelligence, based on the fact that honeybees are faster at learning colors than other vertebrates, including humans. Insects have surprisingly large behavioral repertoires given their small brains, with flexibility that rivals that of at least some vertebrates. Maybe, they suggest, the question should be not how insects can do so much with their tiny brains, but why so many vertebrates bother with big ones. Rather than focusing on an All-Animal IQ test, it seems more interesting to focus on the way that evolution has acted in similar ways in distantly related groups.

Rules of Thumb for Creatures That Don't Have Any

I have been studying crickets for over three decades, and I would never suggest that they are geniuses. They can exhibit some rather astonishingly flexible behavior, but they do so in a way that obviates the need for complex cognition, the same way that many other animals might be able to combine genes and experience to produce sophisticated responses to the environment.

The crickets I study, sometimes called Pacific field crickets, live in subtropical parts of Australasia, including northern Australia and islands like those in Samoa, Fiji, and Tahiti. Sometime before the late nineteenth century, people introduced the crickets to Hawaii, where they live ordinary cricket lives: males sing by rubbing their wings together to attract females, and females choose males based on the quality of their songs. In Hawaii, unlike other parts of their range, however, male crickets face another problem besides the one of sounding alluring to a potential mate. A parasitic fly called *Ormia ochracea* (it has no common name) can hear the male's song as well. When a female fly locates a calling male cricket, she flies close to him and drops her tiny, sticky larvae on and around his body. The larvae burrow inside the cricket and proceed to consume him while he is still alive, eating what a student of mine referred to as the "gooey bits" until, a week or so later, the cricket is a hollow husk as the larvae emerge to complete their life cycle.

While arresting, not to mention gruesome, in many respects, this situation poses a conundrum for the crickets. The more a male sings, the better his chances of attracting a mate, but in Hawaii, he also risks attracting the deadly fly. The evolutionary way out of this conundrum has occupied my lab for the past many years, but a relatively recent development was also one of the most startling:[30] on some of the islands where the crickets and flies occur, cricket males evolved a change in their wings that makes them unable to sing. A single gene caused this alteration, and males with the new form—we call them flatwings because the wings lack the structures needed to sing—are now prevalent in several places. The change came about in only a handful of generations, making it an example of extremely rapid evolution in the wild.

Being a flatwing cricket is obviously helpful to males in protecting them from the parasitic flies, which cannot find silent prey, but it also poses obvious drawbacks to the other side of the equation, the attraction of females. Yet the mutation was able to spread, and flatwings—as well as a stable handful of crickets lacking the mutation—are still present in Hawaii. How are the mutant males managing to find females?

Since some of the males can still sing, we wondered if the males with the mutation were behaving as what are called satellites—males that do not signal to females, but instead lurk near the singers, capitalizing on females that are attracted to the singers and mating with them as they are encountered. Such tactics are reasonably common in animals, with frogs, birds, and other cricket species all having satellite males that take advantage of the hard work of other individuals.

A set of field experiments showed that this was exactly the case.[31] The design of the experiment was simple: put out speakers playing prerecorded cricket song, and see who showed up. Flatwing males were more attracted to the song, and they got much closer to the speaker, than the males that had normal wings. But how did the males know, as it were, to do that? The single gene that affected wing development couldn't possibly also affect the brain in such a way that males could know they couldn't sing, then figure out to approach a caller, and then calculate their position. As I said, crickets just aren't that smart.

Instead, we think that the crickets employ what you might call a rule of thumb, a simple algorithm that links a set of environmental circumstances with a preexisting response. A world with flatwings is also a world that is silent, at least from the standpoint of cricket song. But silent worlds can arise for a number of reasons: a recent storm wiped out much of the population, the habitat is poor, or the lawn (which is where the crickets occur) just got mowed. What if the crickets already have a rule that says, in effect, "If I do not hear a lot of crickets singing, I should move closer to the ones I do hear." This does not require the males to actually think that, the way a human would, just to execute a simple If-A-then-B series of behaviors. If they already have that flexibility of behavior, then the mutation could get a head start.

That is exactly what the crickets seem to do. In my laboratory, we have populations of the crickets from many places. We can rear them with and without cricket song, simply by playing recorded song inside the incubators where we keep them. Male crickets that grow up hearing a lot of crickets singing, which presumably sends the message that mates are likely to be available, are relatively slow in responding to a playback of song as adults.[32] The males that grow up in silence, however, navigate to the source of the playback with alacrity, which in nature would increase the likelihood of coming across a female cricket likewise moving toward the song. Females seem to use similar rules, which also makes them more likely to mate with one of the silent males if they encounter one. Note that the system depends on the persistence of the singing males, because they are the linchpin that holds the system together. Every time we go to Hawaii to see what the crickets are up to, I brace myself to find that the populations have gone extinct because too many flies have discovered the singers. So far, that hasn't happened.

Similar rules of thumb may apply to many other species, in which a combination of genes and environmental influences can produce remarkably complex behavior that looks as though it was the result of human-style reasoning. Butterflies, for instance, have to choose a plant on which to lay their eggs, and the quality of that plant makes a big difference to the success of their offspring. My colleague Emilie Snell-Rood and her postdoctoral advisor Dan Papaj did an experiment with the common cabbage white butterfly that showed that females could learn to lay eggs on a plant with either green or red leaves.[33] Green leaves are more common in nature, and the butterflies seemed to have an innate bias to lay eggs on them rather than red plants. Having that bias—what the scientists called an educated guess, but what we could also think of as a rule of thumb—may help the butterflies get a head start in the wild, because the operating costs of maintaining a brain that evaluates red and green from scratch would be too high.

The crickets and butterflies show how flexible invertebrate behavior can be, and that maintaining that flexibility comes at a price. They, along with the crabs, spiders, and octopuses, also flatly contradict Macphail's idea that I mentioned in the previous chapter, that no differences in intelligence exist among nonhuman species. Macphail was talking about vertebrates, but a group of psychologists published a paper[34] in 2020 suggesting that a species of jumping spider—the kind that doesn't build a web, but instead pounces on prey that cross its path—would be a good candidate for an intelligent arthropod whose behavior could counter Macphail.

They detail the abilities of the spider to solve puzzles; change its behavior if it is offered different types of prey; "count," or at least assess rough quantities of objects, about as well as a human infant; and generally behave in what one might call an intelligent manner, at least if we use the same measurements on spiders that we do on birds or mammals.

This all seems well and good to me, but I remain puzzled at anyone taking Macphail's suggestion seriously for even a moment. Had he never seen different animals behaving in the world? And even if he had not, believing his hypothesis would require placing behavior in a completely different universe from the rest of the animal, since no one would ever argue that animals do not differ in physical attributes. Why would all animals have the same learning ability, or however one might define intelligence, but have different digestive systems, eyesight, or numbers of limbs? It doesn't do us any good to separate the two, as I have been pointing out.

Putting the Joy Back in Killjoy

Even though animals can do extraordinarily complicated things without the same kind of cognition that humans have, we still seem to want to categorize them by their similarity to us humans. Indeed, the popular press seizes new evidence of tool use, problem-solving, or social plotting in animals as evidence that other species are smarter than we had previously thought. Many of the comments from readers on such coverage seem to take a kind of vindicated tone, as if science is finally catching up to what they had known all along, that animals have evolved the exact same abilities and inner lives as people and we are being needlessly closed-minded by failing to recognize that fact.

Not everyone, however, is convinced. Sara Shettleworth, a comparative psychologist at the University of Toronto, wrote a paper[35] in 2010 lamenting our dismissal of simpler explanations for complex behavior as "uninteresting and 'killjoy,' almost a denial of mental continuity between other species and humans." This use of the term *killjoy* comes from Daniel Dennett, a philosopher who has written about the evolution of consciousness, a similar problem. Shettleworth goes on to point out that although we have a kind of folk psychology that we use to explain what animals do, that intuition is often simply wrong. It is even wrong when we apply those explanations to our own behavior by assuming that people are making conscious decisions all of the time. Psychologists have known for many decades that less rational mechanisms can account for what we do, from making bad economic choices to deciding whether to put money into the coffee kitty at work—a study showed that the image of a pair of eyes "watching" contributors increased the amount of money left to more than twice what they donated if the image was of a flower.

Such tendencies to see animals as exactly like people are hardly new. In 1898, the eminent psychologist Edward Thorndike published what can only be termed a rant against such anthropomorphism,[36] saying, "The history of books on animals' minds thus furnishes an illustration of the well-nigh universal tendency in human nature to find the marvelous wherever it can." He laments that "folk are as a matter of fact eager to find intelligence in animals. They like to. And when the animal observed is a pet belonging to them or their friends, or when the story is one that has been told as a story to entertain, further complications are introduced." It boggles the mind to imagine what Thorndike would have had to say about internet cat memes.

This doesn't mean that we need to go back to human exceptionalism. Daniel Dennett has tried to call for a compromise,[37] saying, "People in the field often gravitate into two camps. There are the romantics, and the killjoys. I think the truth is almost always in the middle." But this is a case where it's hard to see what the middle means here. That one should be slightly romantic, or that animals are somewhat like us, but not exactly? Like many attempts to find

middle ground, doing so seems prone to leaving both sides unsat-
isfied. In his discussion of consciousness, Dennett also criticizes
an earlier writer for suggesting that consciousness "presumably"
occurs in mammals, and "probably" in birds, reptiles, and amphib-
ians. "Wondering whether it is 'probable' that all mammals have *it*
[consciousness] thus begins to look like wondering whether or not
any birds are *wise* or reptiles have *gumption*: a case of overworking
a term from folk psychology that has lost its utility along with its
hard edges."

A similar objection could be made to the idea that all animals
have the same intelligence or emotions as people. Invertebrates particularly reveal the shaky ground of that overworking, because if it
seems a bit odd to talk about a turtle with gumption, it is manifold
weirder to speak of a mournful caterpillar or a wistful snail. Some
authors say that the relationship between animal intelligence, or
consciousness, and our own is more of a continuum, with shades
of gray rather than a sharp us versus them distinction; Godfrey-
Smith, for example, points out that being conscious is not an on-off
switch, and thinks of it as more of a graded experience.[38] But this,
too, seems less than satisfying, because a continuum is still a line
marching from one shade to the next, with white at one end and
black at the other, like the scala naturae we know to be incorrect.
Intelligence doesn't grade from fish to turtles to birds and finally
primates, so it doesn't really help to suggest that the edges between
them are blurry. Not having an on-off switch doesn't mean that we
must perforce have behavior on a dimmer.

What does help, once again, is thinking of behavior as evolving
the same way that other characteristics do. No one tries to pit killjoy versus romantic explanations for the differences among, say,
kidneys in different kinds of animals. We can see that our kidneys
are more similar to those of other mammals than of birds, and all
three groups are distinct from insects, which use a completely different system called the Malpighian tubules to filter waste. Yet no
one worries about who should be admitted to the urinary process-

ing club of True Kidneys. Animals differ in many attributes, including behavior, but we only seem to want to create rankings for some of the attributes.

Finally, the real prize for killjoy approaches to animal behavior must come from a group of psychologists who published a paper cheerily titled "Towards an Animal Model of Callousness."[39] While they take no position on whether animals are callous in the wild, the authors suggest that because a lack of empathy—callousness— in humans is an important antisocial disorder, maybe we should find a relevant callous animal model to explore, the way we have with other mental disorders. Research on empathy and callousness in humans shows that callous people differ from more empathetic ones in several parts of the brain, particularly in part of the frontal cortex and the amygdala. This link to brain function suggests that exploring dysfunction in those pathways might yield insight into the problematic behavior. A slight difficulty seems to be that rodents—their proposed choice—do not ordinarily exhibit anything like human callousness, and even prefer situations in which they and a companion get a reward over getting the same reward by themselves. Not daunted, the researchers speculate that if that generous impulse could be eliminated, either with a different experimental setup or with drugs, it would be possible to examine its opposite in a controlled manner. Whether finding a rat that wouldn't sympathize with your problems would be a help in our considerations of how we perceive other animals' abilities is another question.

9

Talking with the Birds and the Bees. And the Monkeys.:
Animal Language

I n one of the first signs that my then new husband trusted me, he agreed, albeit reluctantly, to let me take William the cat to Albuquerque. I was beginning my postdoctoral work at the University of New Mexico, and he was staying behind at his faculty job in Ohio before joining me in the summer. I thought William would be a good companion as I started out on my own, though the cat had been my husband's before we got together. Since I would take the cat with me when I flew down, I figured it would be a good idea to get him some kind of sedative to keep him calm during the journey. I duly inquired at the vet's office about what they could prescribe. The veterinary technician looked thoughtful. "Is the cat very verbal?" she asked.

Verbal? William produced the usual assortment of meows and purrs, but had never actually, well, talked. I suddenly had a sense of inferiority, wondering if other people had had talking cats all this time and mine was just inadequate. Then I realized that of course she meant vocal, not verbal, and I heaved an internal sigh of relief. No, I assured her, he wasn't particularly prone to making noise. The tech proceeded to give me some pills for William, and he remained suitably subdued during almost the entire trip, waking up to express his displeasure just as the plane touched down in New Mexico and yowling nonstop down the jetway, through baggage claim, and into the taxi home. He never, however, uttered a word.

Animals being verbal, and not merely vocal, is not just a matter of concern for those of us with a sudden fit of pedantry in the veterinarian's office. Language is often viewed as the final wall between humans and other animals, and a characteristic whose evolution seems obscure. Even after tool use fell to the bees and crows, and the mirror test became something fish could pass, the Rubicon that remained was language. Raymond Bergner, the psychologist who fretted about our inability to define behavior mused:[1]

What is different with respect to animal behavior is not the parameters that apply but the capacities of infrahuman species. For example, aside from the modest sign-linguistic ability of certain species such as chimpanzees, the great majority of animals seem to possess neither language nor a capacity to acquire it. Accordingly, they are not able to act on the enormous range of discriminations that are captured in human language and that are thereby available to human beings in their behavior.

After undergoing a religious conversion, author Andrew Norman Wilson became skeptical about the evolution of behavior, particularly in humans. From a very different starting point than Bergner, he comes to a similar conclusion:[2] "Do materialists really think that language just "evolved," like finches' beaks, or have they simply never thought about the matter rationally? Where's the evidence? How could it come about that human beings all agreed that particular grunts carried particular connotations? . . . No, the existence of language is one of the many phenomena—of which love and music are the two strongest—which suggest that human beings are very much more than collections of meat."

In a more scholarly vein, although they generally agree that indeed, language "just 'evolved,'" linguists, psychologists, and biologists have spent decades debating, often with an astonishing degree of acrimony, even for academics, whether it arose suddenly in our evolutionary history or more gradually. If the latter, what are its best counterparts in other animals, and which qualifies as similar to us? According to the view that language arose suddenly, humans—in particular, the behavior of humans, not their appearance or bipedalism—are distinct. That means if we want to settle the arguments about the inherent brutality of men that I brought up in the introduction, we are stuck with those brutes, because some behaviors, at least, evolved in a way unlike other characteristics. In a 2003 article[3] in the *New York Times*, Nicholas Wade claims, "The only major talent unique to humans is language." Whether language is a talent, akin to playing the violin or shooting hoops, seems questionable, but language seems to have remained as the last bastion of human behavior categorized as being special, that is, sudden. As should be apparent by now, I think this argument is false. Behavior, even a behavior as complex as language, evolves the same way that physical appearance does. This doesn't mean that human language lacks unique components, but that those components are not somehow outside the rules governing the evolution of other traits.

This chapter takes a critical look at claims about who talks and who doesn't. What constitutes language? Is it the same thing as communication, putting language on a par with screeching monkeys or mimicking parrots? Does it require what some researchers call mental time travel, the ability to imagine oneself in another time or place? And is language required for thought itself?

We Don't Want to Talk about It

Although people have been speculating for millennia about where language came from and how it differs among cultures around the world, modern scholarly consideration of the origin and evolution of language got off to a rocky start when the Linguistic Society of Paris banned discussions of the beginning of language in 1866 because "it was an insoluble metaphysical problem."[4] Leaving aside that being insoluble, much less metaphysical, doesn't seem to stop people, particularly academics, from discussing other topics, the society had a point. As is often said, behavior doesn't fossilize, and we have no exact analogues of language among animals, though there are some interesting approximations, as I will discuss later. At the same time, the nearly seven thousand human languages of the world seem almost eerily similar: no human society has a language that can be described as less fully developed in its structure or its use of symbols than any other language. We therefore don't have an obvious intellectual foothold to determine how language itself began, because there is no linguistic equivalent of trying to understand the eye by examining simpler light-sensitive organs in an invertebrate ancestor. How did we get to modern speech from primordial ancestors that lacked it?

Some evolutionary biologists see human language as representing one of the "major evolutionary transitions"—steps that made a huge difference in the way that life on Earth proceeded. Some of the other transitions include becoming multicellular (composed of many interacting cells instead of a unit with a single nucleus), and reproducing sexually, with genes from two individuals combining rather than being replicated as an identical copy. Like those other steps, language opened the way for many other abilities. Being able to communicate with symbols, to refer to things in the past and future as well as the present, means that people could form complicated social groups, with alliances that had reference to previous actions instead of only immediate responses.

In *The Goodness Paradox*, anthropologist Richard Wrangham suggests[5] that the crucial function of language is that it allows gossip—the discussion of others' behavior. That in turn means that someone's reputation is made and can be used as a basis for social decisions—who gets rewarded? Who gets punished? Wrangham is particularly interested in the latter question, because a society that can impose moral sanctions based on prior behavior can ensure that delayed punishment is meted out. Chimpanzees, in contrast to people, do not seem to care what others think; an experiment in which one chimp sometimes could see another individual steal food from a third didn't make any difference to the thief's behavior. The same thing applied if the chimps could see another individual being helpful. In humans, of course, language allows people to explain their motives, deal with mistaken impressions, post inflammatory material on Twitter, and coordinate punishment, up to and including execution of those who violate social contracts. According to Wrangham, no human ability is "as special as the one that enabled conspirators to trust one another sufficiently to collaborate to kill a bully."[6]

Whether or not one accepts coordinated and preplanned punishment as one of its primary outcomes, let alone benefits, language has made a huge difference in human evolution. Some scholars see the acquisition of language as part of a number of evolutionary innovations that allowed humans to become humans. These attributes include bipedalism, which in turn allowed early humans to carry objects in their hands rather than their mouths. Carrying things in one's hands then paved the way for language, according to a paper[7] by Charles Hockett and Robert Ascher published in 1964, because this freed the mouth for other activities. "What were the hominids to do with their mouths, rendered thus relatively idle except when they were eating? The answer is: they chattered."

No one else seems to have picked up on this hypothesis of speech-as-side-benefit-of-hands-free-living. Hockett and Ascher also segue[8] into the unsettling idea that speaking with the mouth isn't required, and that if evolution had proceeded differently "speech sounds today might be anal spirants," which are exactly what you think they are. That no other animals communicate via this channel did not seem to discourage them, but the notion does not seem to have gained much traction, which is probably all to the good.

As Old as Thought Itself?

The ban by the Linguistic Society of Paris has now been lifted, of course, and fraught discussions of the origin of language abound. Some scholars still maintain that language emerged all at once, and is a universe apart from communication in other species, while others see a more gradual development. As I also noted, I fall into the latter camp. Scholars' thinking about this issue has been helped by developments on two fronts.

First is what we mean by fast evolution. Although Darwin emphasized the slow, incremental pace of natural selection in producing evolutionary change, a lot has happened since his time, and scientists now recognize that evolution—changes in the proportions of different genes in populations—can happen extremely quickly. The crickets that I study in Hawaii evolved that silent male form in fewer than twenty generations, and that's not even a record. Admittedly, losing the ability to sing in an insect is far different from gaining the ability to not only sing but also recite poetry and rant about politics, but many, many examples of rapid evolution have been recorded. Microbes, of course, are the champions of speedy genetic change, and insects are not far behind, but fish, birds, and even humans can respond remarkably quickly to pressure from the environment. Guppies, for example, evolve to become sexually mature at an earlier age in places where their predators abound, which helps them reproduce before they get eaten; this is a genetic change in the way they develop, not a response to immediate conditions.

Part of the difficulty in thinking about language evolving is that it can be hard to understand how any complexity, language included, can arise from small changes, each building on the one before. Andrew Norman Wilson's bewilderment[9] about all human beings having "agreed that particular grunts carried particular connotations" may stem from a misunderstanding of the way evolution proceeds. That misunderstanding is common, and it is related to the widespread belief I have mentioned before, that behavior is different from physical attributes. Humans didn't get together and agree on grunts having meaning any more than vervet monkeys discussed which call means "predator from above" and which means "predator from below," though such differentiation is well established. For that matter, mammals didn't get together and discuss how fur would be structured, or vertebrates the way that the immune system would recognize foreign substances and manufacture antibodies to remove them from the body. But just because something is complicated doesn't make it impossible. Language evolved like other complex mental, and physical, characteristics.

The second development in how we look at language is in the ability to detect its precursors in ways that the Paris linguists probably couldn't even have imagined. The fossil record was extremely limited in 1866, and back then we knew much less than we do today about the ancestors of humans. Many scholars think that language emerged perhaps one hundred thousand years ago, but a few anthropologists now suggest that language is detectable at least one million years ago, in our ancestors *Homo erectus*, and that it is also possible that Neanderthals had some form of symbolic communication, though this is speculative. The evidence, they say, comes from the use of tools such as hand axes that several people often used for a long time or hoarded with other items, a behavior that indicates a more sophisticated form of communicating than can be achieved with gestures or a few vocalizations. Long-distance ocean voyages, undertaken by people in islands in various parts of the world, also suggest that people had to discuss events in the future and coordinate their activity, again difficult to do without symbolic communication that can refer to events occurring in something other than the here and now. Symbols—items with meaning beyond their physical existence—do not just occur in rituals such as burials; they can be part of our day-to-day life. Thus, if people had a tool with an agreed-upon use, that might count as a symbol. What exactly constituted such early forms of language is, of course, unknown.

Another source of evidence for a long-ago and gradual origin of language is physical, but of a very different nature. No other primate besides the human speaks, and for many years scientists assumed that nonhuman primates were physically incapable of even approximating human language because of limitations in their mouth, throat, and brain. This apparent gap meant that at least our kind of language really did seem to have appeared all of a piece, without much in the way of ancestral foreshadowing in non-human primates.

Now, however, that assumption is being questioned. A group of European and US scientists performed some highly sophisticated brain imaging on a macaque,[10] a kind of monkey that has not shared a common ancestor with humans for twenty-five million years. They found a structure that forms a pathway between the auditory part of the brain and the frontal lobe, similar to the component that is seen in humans and that is part of the machinery needed for speech. This link suggests that the seeds of human language are buried deep in the primate lineage. Even though the macaques cannot speak, the ancient origin of the pathway means that human language has come from the same kind of parts-lying-around-in-the-garage that other characteristics do.

Another anatomical theory about human speech and its uniqueness that has fallen in recent years is that of the descended larynx. The larynx, sometimes called the voice box, is a structure that sits in the neck and is important for both breathing and speech. Other mammals have a larynx, including nonhuman primates, but some scientists had thought that because of its position high in the throat, it was simply impossible for those animals to produce speechlike sounds, particularly vowels. In humans, the larynx descended—moved to a lower position in the throat—around two hundred thousand years ago, which used to be thought of as the absolute floor for the evolution of language. And indeed, efforts to try to teach chimpanzees, for example, to speak had resulted in failure, though apes, like the famous gorilla Koko, could learn to communicate with humans using symbols.

But more recent examination both of the anatomy of monkeys and apes and of the necessity for certain structures in producing vowels tells a different story. That descended larynx isn't unique to humans after all. It is seen in other animals, including some primates. And those primates do produce calls in a way that is similar to the way people make vowel sounds, in a manner called "speech-ready" by some researchers. Interestingly, some of the best examples of speechlike sounds come from marmosets, tiny monkeys that are much less closely related to us than are the apes. Like humans, as marmoset infants get older, they change their sounds in a way that reflects their brain development. So speech may have its roots much further back than two hundred thousand years, just as the brain imaging studies suggest. The authors of the anatomical study conclude that "the idea of recent, sudden, and simultaneous emerging of speech and language is no longer plausible."[11]

Finally, people sometimes derisively refer to talking as "flapping your lips," which is a vivid if crude description. Scientists being what they are have measured the rate at which those lips are flapping, or more accurately, the number of times per second someone opens and closes their mouth. This turns out to be remarkably standard across all spoken languages, at between two and seven open-close cycles per second. In chimpanzees, which smack their lips to communicate, that rate is four cycles per second, and other apes are similar to them. This means that speech could have arisen from what a 2020 paper[12] calls "ancient primate rhythmic signals," which is a rather more dignified phrasing than lip flapping.

Monkey See, Monkey Point, Monkey Say

So if language did not emerge full-blown, all at once, then what did the first speech sound like? One possibility is that it was similar to pidgin, a term for a simple cobbled-together language that modern people devise when they do not share a language. For example, people on Pacific islands who come from a number of different places, including English-speaking countries, may speak pidgin English when they need to conduct business together. Such simplified languages may have arisen before the more grammatically complex tongues that are spoken today.

Another option is that gestures preceded speech, since spoken language is just one form of communication. Michael Corballis, who was a professor of psychology at the University of Auckland, suggested exactly that,[13] and also focused on "mental time travel," the ability to imagine what might happen in the future and what has happened in the past, as a crucial part of the evolution of language. Being able to refer to what he called "the nonpresent" is particularly important. This ability, he argued, while clearly well developed in humans, may also be seen in other animals, like jays that remember where they stashed nuts and return to the locations with precision much later. Corballis was skeptical of the idea that language appeared suddenly, and noted that trying to separate human language completely from anything animals do "can also lead to overly simple explanations for animal behavior, leading to smug superiority and the invocation of what seem to be miracles."

Other animals, of course, use gestures, and both of our closest relatives, the chimpanzees and bonobos, have a rich repertoire of motions that seem to be shared at least with each other. A 2018 study[14] characterized both species' gestures, and tried to assign meanings to each distinct movement or set of movements using something called the "'Apparently Satisfactory Outcome' (ASO), the reaction of the recipient that satisfies the signaler as shown by cessation of gesturing." In other words, if you wave at someone across the street and they wave back and you then stop waving, it's an ASO. One is tempted to think of scenarios in which gestures are not followed by signaler satisfaction, per se, with communications among drivers in traffic coming to mind, but the idea is still clear. It turns out that chimpanzees in the wild have at least nineteen ASOs, and their repertoire has about a 90 percent overlap with that of bonobos. The gestures not only look similar but also seem to mean much the same thing, which in principle suggests that the two species could understand each other.

What about people? We often focus on other animals' abilities to interpret our language, but how about our ability to interpret theirs? The Great Ape Dictionary[15] project aims to answer that question by having people view videos of chimps and bonobos performing a gesture, maybe reaching an arm out or doing what the project leaders call a Big Loud Scratch. Led by primatologist Richard Byrne of the University of St Andrews in Scotland, the project enlisted the aid of thousands of people watching twenty videos of chimps and bonobos and choosing which interpretation they think fits the apes' gestures the best. The data are still being analyzed, but the possibility of a kind of universal primate gesticulation, one that points to (pun intended) a very ancient common evolutionary history of communication, is appealing.

The analogy between ape gestures and human speech only goes so far, however. Human infants gesticulate before they can speak, and baby sign language has become popular among some parents eager to interpret the arm waving of their children long before it is realistic to ask them to "use their words." In humans, those gestures affect the way a baby's cognition develops as he or she grows up. But a careful comparison of ape and human infant gestures[16] shows that some of the similarities at least are superficial; the nonhuman primates do not point at external objects the same way that babies do, and their gestures do not show the same kind of cultural changes over time. This does not mean that ape gestures are inflexible—they can be adapted to circumstance the way that human gestures can, and they can show intentionality, which means that the gestures are about something, rather than simply expressing emotion. They do not, though, contain all the elements of language.

It's (Not) Only Human

Even with all of those connections between human speech and animal signals, no one would argue that animal communication is exactly the same as language. The Linguistic Society of America puts it even more starkly on its website:[17] "No other natural communication system is like human language." That depends on what you mean by "like," since many species communicate in ways that are at least somewhat similar to the way that humans do, even if those communications are not identical to our own speech. Perhaps Sara Shettleworth put it best: "The old question, 'Can animals learn language?' has been replaced by appreciation that although human language is just that, *human*, other species share important components of it."[18]

Part of the reason that people have seen such a gulf between human language and animal communication is that we have tended, for both good reasons and bad, to focus on primates. The good reasons include our recent common ancestry, which makes apes and monkeys an obvious choice to look for common structures or neural patterns that could have led to language. The bad reasons are similar, because as is the case with cognition more broadly, as I discussed previously, convergent evolution can mean that distantly related animals independently evolve similar characteristics, and the way those characteristics got there can be illuminating.

Truth be told, primates are a bit disappointing when it comes to language. They aren't musical like birds or even whales, they don't seem to have individual names for each other like dolphins, and while they have complex gestures that may even be interpretable by other species, those gestures mean rather ordinary things like "let's have sex" or "climb on me." I suppose one could argue that the vast majority of human communication is equally prosaic, but one might have hoped for more from our closest relatives.

Instead, we can learn a great deal about language and how it might have evolved in humans by considering a wide range of other creatures. Animals signal to each other with smells, sounds, and sights, and often do so with an impressive degree of complexity. That said, the link between having a lot of signals and animals' cognition is not necessarily straightforward; we tend to assume that an animal with more complex communication must be smarter, but that is our anthropocentrism talking. Like humans, animals seem to know more than they can express. For example, a honeybee can remember many different colors, and can even see in the ultraviolet range, but a worker's waggle dance, a complex form of communication with at least some aspects of language, only includes information about the location and distance of flowers, and not their color. Humans, of course, also have many concepts that go unexpressed (for which we can all be grateful, since presumably the alternative is an endless narrative of what is going on in all of our heads), and animals may also know more than they can say, as it were.

The idea that one can have concepts without the words to express them leads us to an often-considered problem in the evolution of language, namely whether language is necessary for thought to occur at all. Darwin believed that language was necessary for any kind of complex thinking and hence complex behavior, and assumed that animals lacked either one. But he was unaware of many of the impressive accomplishments of animals such as the New Caledonian Crows or even chimpanzees, since few detailed field or laboratory observations of behavior had been made at the time he was writing. Because we humans find that language helps us think, and because it is inextricably tied to our imaginations, it is hard to come up with a definition of thought that doesn't require it. Peter Godfrey-Smith questions the idea that thought requires language, pointing out that the aforementioned birds do what can only be seen as complex internal processing of events. Language, he says, "is not essential to the organization of ideas, and language is not *the* medium of complex thought."[19] As Godfrey-Smith emphasizes, just because we do something a particular way doesn't mean it is the only way. If you always assume things happen the same way in other animals as they happen in people, you end up with a circular and not very interesting definition of whatever you are trying to explain.

None of this is to say that bees can talk, or that dolphins, even though they can master concepts like "press the lever on the right to get food," are thinking things like "I wonder when that stupid keeper will give me a test that lets me show off my real skills." But it does mean that, as Corballis suggested,[20] "The burden of explaining language evolution is lessened . . . if language is regarded as communication, not thought."

Vocal Learning: If You Hear Something, Say Something

One of the hallmarks of language, and one that we share with at least some animals, is vocal learning. Infants are born crying, not declaiming Shakespeare, and take years to get from the former to the latter through a process that is a fascinating combination of genetic input and spongelike absorption from the environment. The jargon term for what the babies are doing is *vocal production learning*, which means the ability to make a sound based on what you hear, or change those sounds based on social feedback. It is found in three groups of birds: songbirds, parrots, and hummingbirds—and five groups of mammals: humans (but not other nonhuman primates), dolphins, bats, elephants, and seals. This somewhat erratic distribution, and the length of evolutionary time since the groups shared a common ancestor, means that vocal production learning arose more than once, buttressing the argument that complex communication is the result of convergent evolution. (A broader kind of vocal learning, auditory comprehension learning, means that the listener can learn what a particular signal conveys, and is the kind of learning found in dogs and other animals that respond to verbal commands.)

In birds, we are probably most familiar with vocal learning in the form of mimicry, with parrots and a few other species like mockingbirds being able to imitate a wide range of sounds by their own and other species, including human speech. The many birds that learn their own species' songs are perfect examples of the exquisite interplay of innate predilections and learning. The degree to which each of the two is combined differs among birds, even closely related ones; White-crowned Sparrows, for example, need to hear a White-crowned Sparrow song during a critical period while they are in the nest in order to produce the correct song later, while Chipping Sparrows just have to be able to hear themselves sing and do not need a tutor.

One of the most remarkable aspects to songbird learning is that the young bird hears its song months before it produces it on its own, and even then goes through a period of refinement after it starts to sing. I can always tell when the juvenile White-crowned Sparrows have arrived in the spring because they seem to have a slightly screechy quality to the song, like a ten-year-old enthusiastically belting out a school anthem off-key. This learning-plus-genes process also means that in some species, including the white-crowns, so-called regional dialects can develop. A professor of mine, Barbara DeWolfe, was one of the pioneers in birdsong study, and she could walk through the campus at the University of California, Santa Barbara, where I was an undergraduate, and tell us where a given bird had come from just by listening to him sing. "San Francisco," she would say, or "Oregon," an ability I still think seems nothing short of magical.

Why vocal learning occurs in some animals and not others is not fully understood. The distinction isn't binary—what you might call partial vocal learning is seen in a range of species, including, somewhat surprisingly, mice, which can tailor their vocalizations in the lab with training but do not seem to do so in the wild. We also do not know the function of full versus partial vocal learning: What do the Chipping Sparrows have that the white-crowns lack, or vice versa? Biologists have a number of theories, but much remains to be done. Even in species with robust vocal learning, some aspects of response to vocalizations are probably not learned—having to find out the hard way that a particular call means "a hawk is about to descend" wouldn't be very practical, to say the least.

A somewhat unlikely candidate for future research on vocal learning was proposed in a 2019 paper[21] by Sonja Vernes and Gerry Wilkinson that championed bats as a good model, at least for mammals. Most people think of bats, if they think of them at all, as fairly silent and solitary animals, but many species are quite social and will roost in large groups during the day. I once saw a tree in Australia that was festooned with fruit bats, sometimes called flying foxes. The bats were hanging upside down, more or less shoulder to shoulder, or what passes for that in a winged creature, and the sound was unrelenting: squeaks, chirps, and grunts as the bats shuffled for position. I can't imagine that it was easy to sleep in the cacophony, and I fancied that if they could have spoken, it would have been a nonstop litany of "Move! Your wing is in my face!" "No, *you* move—I just need to stretch and your head is in the way!"

Vernes and Wilkinson[22] do not suggest that bats learn to complain, but they do note that the mammals use individualized calls that let mothers find their babies when the adults return to the roost, and they also change their echolocation calls depending on feedback from other bats in their colony. Pups will match their calls to their mothers' calls. In the disc-winged bat, pairs of individuals will produce alternating calls that Vernes and Wilkinson say resemble the swimming pool game of Marco Polo so that they can find each other in the roost. Finally, male bats in some species sing to attract females, which makes their vocalizations seem like the bird songs that are learned by the sparrows and others. A key difference is that unlike songbirds, bat fathers do not stay around the nest or roost while the pups are young, so the babies do not have a similar opportunity to hear their fathers sing, which suggests a stronger genetic influence on bat song. Still, bats may provide an intriguing test case for understanding vocal learning in mammals.

The Babbler's Inadmissible Transgression

If, as the nineteenth-century scholar Friedrich Max Müller put it, "Language is our Rubicon, and no brute will dare to cross it,"[23] at least a part of that impassable boundary is syntax. Syntax refers to the way that the different parts of language are combined into meaningful units, and has long stood as an impenetrable barrier between human and animal communication. Although bird and whale songs have ordered components in them, and even my crickets sing with distinctively repeated elements, these are not generally viewed as containing syntax because the units themselves lack independent significance and cannot be recombined to have different meanings.

As with so many other barriers, however, the one between human and animal syntax may also be permeable. Some birds, such as Japanese Tits and Southern Pied Babblers, produce calls that can in fact be recombined. The tits have "alert" calls that warn their neighbors about potential predators as well as "recruitment" calls that tell others to come to a good food source. Unlike many other birds, they can combine the two calls into one that means, in effect, "come over here and help me attack this predator," which qualifies as syntax at least according to some scientists. The tits can even respond to the calls of a related species, the Willow Tit, in a way that suggests that the birds can assess the meaning of individual components rather than just hearing a generalized "Hey, a little help over here!" message. Other scientists are pickier, and according to a 2019 article,[24] "perceive this [animals having syntax] as an inadmissible transgression."

A novel approach to determining whether birds or other animals might have language-like components in their signals was undertaken by a group of researchers using a method that is generally applied to understanding species diversity. When we speak of the tropics as having a greater biodiversity than, say, the Arctic, we mean two things. First is the number of species: the Arctic has about 200 kinds of breeding birds, for instance, while Brazil has over 1,800. But we also want to take into account the relative abundance of those species, since if each of the 200 in the Artic had a population that was ten times the size of those in Brazil (they do not, but let's say so for the sake of argument), that would make the Arctic comparably diverse at least in one sense of the word. Estimating those numbers is difficult, and doing so requires the use of sophisticated mathematical models.

The researchers, based in the United Kingdom, the United States, Israel, and Germany, applied those models to language,[25] but instead of species number and abundance, they used the units of communication—components of either human speech or animal signals. It turns out that the information content of all human languages follows a mathematical rule, called Zipf's law, which is in keeping with all languages also being able to express the same kinds of ideas. But it also turns out that the units of communication in songbirds, rock hyraxes (small mammals that are related, improbably, to elephants), and dolphins follow Zipf's law as well, showing that the patterning of human language is roughly similar to that of the other animals.

A concept that is related to syntax, and is also a characteristic of human language, is called combinatorial signal processing, which means that smaller elements are grouped into larger ones, like syllables into words and words into sentences, in an ordered manner. Again, this is often thought to occur only in humans, but a very similar ability was recently discovered in a treehopper. Tree-hoppers are insects that look a bit like they have dropped in from outer space; oddly shaped, many of them also sport elaborate head-gear, including one species named after Lady Gaga (*Kaikaia gaga*). The species under question doesn't have quite such extravagant ornamentation, and instead resembles a rose thorn with legs. It even lacks a common name, going by the unprepossessing Latin title *Enchenopa binotata*. Like others of their kind, male *E. binotata* signal to prospective mates by rhythmically vibrating the leaf or stem on which they are perched. The signals are patterned, like tiny drumbeats, and a group of researchers from the University of Wisconsin-Milwaukee,[26] led by Bretta Speck in the laboratory of Rafael Rodriguez, who has been studying the tiny creatures for many years, asked whether the female treehoppers would respond to the signals if the beat, as it were, was altered. In other words, if they usually responded to the equivalent of "Hey, come over here! I am a terrific treehopper!" would it work equally well to broadcast, "Here over terrific! Treehopper hey come am a!"?

The answer was no. Female treehoppers were quite discriminat-ing in their responses, and seemed to have internal rules about how the parts of the signal should be combined. The authors do not claim that treehoppers are talking, or that grammar or syntax in human language arose from the same ancestor in both the insects and people. Instead, combinatorial signal processing, they suggest, "may represent a common solution to the problems presented by complex communication in a complex world."[27]

It is worth noting that just because syntax has only been docu-mented, even controversially, in a few species does not mean that all other species lack it. We simply haven't looked at all of the perhaps ten thousand bird species, for example, or even just the more than six thousand kinds of songbirds, with anywhere near the level of scrutiny and careful experimentation required to establish how they commu-nicate. This points to one of the many problems with declaring that only humans possess this, that, or the other characteristic: How can we possibly know for certain? Remember, the scala naturae is false, so we cannot assume that just because our closest relatives lack syntax, or any other signpost of language, all other species do as well.

At the same time, I am not arguing that all creatures are alike, or that they have the same abilities. Birdsong, complex though it is, does not seem to be related to other cognitive abilities in birds. This means that nightingales, for instance, do not tend to be any better at solving problems using tools than birds with less melodious songs. This contrasts with the situation in humans: people who are verbally skilled also tend to be good at other intellectual tasks. Bird songs also do not refer to objects that are away from the singer, meaning that birds do not have the capacity for mental time travel. Bird songs do show one intriguing similarity with human language, which is that at least some species, including Canebrake Wrens, can take turns while speaking, or, in their case, singing, like humans having a conversation. Usually it is the members of a mated pair that engage in a kind of call-and-response duetting, often with remarkable precision that rivals or even exceeds the way in which humans can intertwine their speech during a conversation. Whether one sex or the other interrupts more remains to be seen.

Like They Know the Score

In the previous section I referred to birdsong as melodious, and indeed the idea that it is like human music is quite popular. For some reason, people do not seem to feel that birds having music is nearly the encroachment on human exceptionalism as the idea that other species have language. Yet both music and language are seen as distinctively human, and theories about how musicality evolved abound, ranging from its being a by-product of evolution for dealing with sound in our environment to it being a sexually selected characteristic, like the flashy tail on a peacock. Just as all human cultures have a language, all of them have some form of music.

So how might birdsong be similar to music? One of the obvious commonalities is vocal learning. Birds can learn complicated rules about sounds—when each note should be produced, how long to wait between segments of a song, and so forth. The ability to memorize patterns in sounds differs among species, and not in ways that are easy to rationalize. For example, Budgerigars, which are a kind of small parrot, can perform some sound discriminations that a Zebra Finch cannot, even though Zebra Finches are part of the group of birds that sing their own complex songs.

Why the variation? Perhaps parrots are just different from other birds, as evidenced not only by their adeptness at mimicking human speech but also by their dance moves. Like Snowball the cockatoo, which I mentioned in chapter 7, Snowball's dance feats spurred a flurry of interest in the idea that being able to move to the rhythms of music also indicated a preadaptation to language. This idea was particularly interesting since, at least in the early research, the species that danced also were good at vocal production learning. That bubble burst, however, when Ronan the sea lion was found to do much the same thing.[28] Humans can teach sea lions to do tricks, but they certainly don't have anything resembling language. So the real question is why some, but not all, of the species that show vocal learning also perceive musical rhythms. This will require more research, but in the meantime, being musical, or at least being able to respond to it, doesn't mean that an animal is on the way to human language.

Doubt has also been cast on whether birdsong is music after all. Marcelo Araya-Salas from New Mexico State University examined a species[29] already thought to be a virtuoso, the Nightingale Wren, a tropical songbird from Mexico and Central America. He analyzed the songs of eighty-one individuals for their adherence to the same rules that appear in human musical scales, and found a very poor match. By human standards, at least, birds don't make music, at least if you want to be a strict constructionist about it. Perhaps, some scholars suggest, we turn the birdsongs into tunes in much the same way we fool ourselves into finding meaning in images in clouds or in hearing foreign languages and thinking we can understand them.

Talk to Dogs and Listen to the Casual Reply

Finally, whether or not animals have language, we have long been fascinated by the idea of talking with them ourselves, a la Dr. Dolittle, the fictional character who could communicate with every animal on Earth. I loved Hugh Lofting's books as a child, though even at a young age I found it perplexing that animals could speak exactly as humans, but that no one else besides the good doctor could communicate with them. And what is more, that he never once thought to simply teach other people to do the same thing (I certainly fantasized about being his pupil and doubt I was the only one). Think of the trouble that would be saved at the veterinarian's office—it would have made my question to the vet about my cat William obsolete. I also was slightly irritated by Gub-Gub the duck, one of the only female characters in the books, being portrayed as fussy and narrow-minded, though I was willing to accept the characterization given that I had never met any ducks myself.

My youthful literal-mindedness aside, whether and how much we can communicate with animals, and how much they understand what we say, has long been of interest for people with pets as well as scientists trying to understand how the brain processes information. I have already mentioned the study[30] in which dogs were placed in an fMRI machine to see how they responded to neutral versus positive words. The dogs did indeed show different brain responses, but that does not mean that they actually interpreted the meaning of the words. Instead, the research suggests common brain functions in response to different kinds of sound, something that was probably at the root of the evolution of human language but says nothing about cross-species understanding.

Another study, conducted by members of the same Hungarian

team that examined the dogs' responses to words, used a technology known as electroencephalography. The technology measures electrical activity in the brain via sensors that are attached to the dogs' heads. They then detected the dogs' responses to familiar instructions, like "sit," and to nonsense words. The dogs' brains responded differently to known words than to different-sounding nonsense words, but didn't seem to distinguish between known words and nonsense words that sounded similar.[31] Whether one of the headlines[32] on a story about the work—"Sorry, Folks—Your Dog Doesn't Really Know What You're Talking About"—was unnecessarily harsh or not, the results reminded me of an old Gary Larson cartoon in which the first panel is captioned, "What we say to dogs," with the owner scolding Ginger in detail for getting into the garbage. The second panel, titled, "What they hear" has a balloon above the dog's head saying, "Blah, blah, blah GINGER blah blah blah."

Larson had a companion cartoon in which a woman took Fluffy the cat to task for shredding the furniture, but this time the balloon over the cat's head was completely blank. But a 2020 study might provide, not a way to stop cats' destructive tendencies, but a glimmer of hope for communicating with them. A group of psychologists in the UK decided to see if the anecdotal stories about cats responding to humans blinking at them would hold up in an experiment.[33] Using twenty-one cats from fourteen households, the researchers allowed each cat to see either a stranger or its owner, and then the person was told to "slow blink" at the cat. In both cases, the cats responded by blinking back, and often approached the person who had blinked at them. The researchers concluded that the cats liked the interaction, and suggested that one could use this form of friendly communication with one's own cats or those encountered on the street, saying, "You can start a sort of conversation."

This discovery notwithstanding, William remained nonverbal his entire life. Maybe I should have tried blinking at him more often.

10

The Faithful Coucal:

Animals, Genes, and Sex Roles

N o matter how changeable or fixed we think our nature might be, nowhere does our idea about how behavior evolves seem more fraught than when it comes to maleness and femaleness. Although I will consider humans in this chapter, I want to focus mainly on animals, since the variety in their sexual behavior can shed light on how our own behavior is shaped. Naturally, then, we will begin by discussing coucals.

I know what you are thinking—coucals are tropical birds that are relatives of cuckoos, and hence you might be forgiven for imagining that, like cuckoos, they are brood parasites and lay their eggs in the nests of other species. This would make them less than ideal for seeing how typical male and female characteristics evolved, since most cuckoos have rather unusual reproductive behavior. As it happens, however, coucals raise their own young like most other birds, so your concern is groundless. Nonetheless, coucals still illustrate why generalizing about animals and sex, and where our notions about gender come from, is more difficult than it might appear.

Even though our understanding of the function of hormones, of the brain, and of genes has gotten better and better, our ideas about the role of biology in male and female behavior is still contentious. Whether it's infamous former Google employee James

Damore asserting[1] that women simply are less biologically inclined to be computer programmers or the *New York Times* op-ed I mentioned in the introduction maintaining that women and men alike are simply at the mercy of the "brutality of the male libido," people want to use biology as a way to defend their position. Whatever the starting point, the argument is the same: "How can you suggest men and women aren't different?" Or, "Boys and girls naturally gravitate toward different things, it doesn't matter how you raise them." Or, "Gender identity is just imposed on people by an oppressive society, and we are all part of a big continuum that we define for ourselves." Nowhere is the supposedly forced choice between learned and innate, nurture or nature, more confused than in matters having to do with sex differences.

Like language, the origin of sex differences is a hill people are willing to die on. But unlike language, where we strive to prove how unlike humans must be to all other animals, with sex differences people sometimes want to emphasize gender roles' inherence. Thus, it is inevitable that boys and girls will play with different toys, men and women will end up in different careers, and both will have vastly different preferences. But why should this be true? Despite these different starting points, because behavior requires input from genes and input from environment, and because behavior evolves like physical characteristics, what is unlikely for language is equally implausible for sex.

Furthermore, when it comes to sex differences and how inherent they are, animals can help resolve the problem, though perhaps not in the way people sometimes think. Those who want to claim the inevitability of male dominance often want to point to animals as examples of how biology dictates our sex roles. But they are not, and for several reasons. First, male dominance is hardly an overarching rule in animals, even those closely related to us. And second, sex- and gender-related behavior is not "hardwired," to use a term I've criticized before, in animals any more than it is in humans, and sometimes even less so in animals. The question, then, isn't why people are different from all other animals in the flexibility of our sex- and gender-related behavior, but how we can understand the extraordinary variation we see among all species, including our own.

And that brings us back to the coucals. For those people who do not know what they look like, coucals are chunky birds about a foot long, with a long tail and a heavy bill. They live in tropical regions of Africa, southern Asia, Australia, and the Solomon Islands. About thirty species have been described, and while they all raise their own young, in some of them the male does most of the work, which is highly unusual in birds. In the Black Coucal, a handsome representative of the group with glossy black- and rust-colored plumage, females do the singing, the defense of the territory, and the establishment of relationships with up to five males at a time, each of which mates with her and takes care of the chicks. In contrast, the White-browed Coucal, which as its name suggests has a raffish white stripe above the eye, is monogamous, and both male and female cooperate to feed the chicks. Male Black Coucals observed by Ignas Safari and Wolfgang Goymann of the Coucal Project in Tanzania[2] fed their chicks twice as often as both members of a White-browed Coucal pair combined, though the young birds seem pretty much the same in terms of their needs and growth rates in both species. In all coucals, the females are larger than the males—a not uncommon pattern among animals—but Black Coucal females are a whopping 70 percent larger, while the White-browed Coucal females are only 13 percent bigger than the males.

The two coucal species live in virtually the same habitat, eating the same diet of insects and small reptiles, and subject to being preyed upon by the same natural enemies. Why, then, did the Black Coucal evolve what is called polyandry, a system in which females are mated with multiple males, while the White-browed Coucal kept with a more conventional share-and-share-alike bird pattern? The short answer is that we don't know, but Safari and

Goymann were able to rule out some obvious possibilities.[3] Of the small number of other bird species with a similar breeding system, like some sandpipers and other water birds, all live with unpredictable food supplies that make it advantageous for the female to mate with a male, leave him to incubate the eggs and raise the resulting chicks, and move on to another mate, all while the food to help her produce those eggs is still available. Yet the coucals are under no such constraints. What is more, all of those other birds have young that can spring up out of the egg and immediately toddle after their parent, like ducklings, needing none of the ferrying of worms to the nest that parents of birds like robins must do. The rationale for polyandry is that when the parents don't have to intensively care for the chicks, the female can be freed up, so to speak, to go off and make more babies. The coucals, however, have the more labor-intensive offspring.

So what is up with the Black Coucals? People often refer to a system in which males do much of the parental care and females compete with each other, rather than the reverse, as sex role reversal. Its counterpart is conventional sex roles, which are what many people think of when they imagine how male and female animals act together: males are dominant, females choose from an array of mates, and females do most of the caring for the young. The coucals, of course, along with sandpipers, jacanas (water birds with extremely long toes, that are sometimes called lily-trotters), and an assortment of others, aren't aware that they are outliers, any more than porcupines, sea urchins, and hedgehogs realize that they are among the few animals with spiny exteriors. Whether it matters to how we humans see animals as rule followers or rule breakers is another story.

It Takes Two

Before we delve further into the question of who has which sex role, whether sex roles really exist, and why, it is important that we consider exactly what we mean by the "sex" part of sex role, and how, or if, that differs from what we call gender in humans. A common distinction between sex and gender is that the former refers to the physical differences between males and females—genitals, chromosomes, and so forth—and gender is defined by the social associations with being masculine or feminine. Over the last several years, the categories have sometimes been mingled, with a number of scholars, not to mention members of the general public, contesting the idea that two sexes exist, or maintaining that both sex and gender exist on a continuum.

Part of the confusion arises from the use of different criteria for classifying males and females. To use humans as an example, if your definition relies on things like "the sex that becomes pregnant," you leave out the many people who never have children, or even leave out, if one wants to be pedantic, women who are past menopause. If you use characteristics like the appearance of genitals, or facial hair, obvious exceptions exist in people who are born with variations of these.

Arriving at a consensus view using such different criteria is obviously impossible. Most biologists who study animals overcome the problem by using the criteria of sex cells—the units that combine to produce a new organism. Males can produce small cells, usually called sperm, and females can produce large cells, called eggs. They differ in other attributes besides size, but importantly, no overlap or intermediates exist—no individuals produce sex cells that are halfway between sperm and eggs. In some species, including some snails and their kin, both kinds of sex cells can be produced in the same individual, but again, the cells themselves are distinct. The evolution of these different forms of cells in sexually reproducing organisms, and why two sexes seems to be a stable state in nature for those organisms, is a fascinating topic in itself, but for my purposes here the distinction is sufficient. Note that males don't have to be producing sperm, or females eggs, at all points in their lives. Other characteristics, like the aforementioned facial hair in humans, show some overlap between males and females. Yet others, such as hip bone and pelvis measurements, can be used to assign sex quite successfully. What this means is that sex isn't binary so much as it is bimodal: the vast majority of individuals are in one category (mode) or the other, with a small number born intersex. In a marvelous 2020 article[4] with the pithy title "Sex Is Real," Australian philosopher Paul Griffiths points out: "Many people assume that *if* there are only two sexes, that means everyone must fall into one of them. But the biological definition of sex doesn't imply that at all." For instance, the aforementioned snails are hermaphrodites, and they have a sex—they may produce sperm at one point in life and eggs at another, or both at the same time—but they cannot be categorized into just one or the other sex the way other animals can. What we don't have is a continuum with no obvious boundary.

It's worth noting that trying to define the sexes in all species by using chromosomes doesn't work either, because the XX and XY chromosomes we are familiar with don't exist in all other animals. Sometimes the outside temperature determines sex, as in crocodiles, turtles, and some other reptiles and fish. Females lay the eggs and then they develop into males or females depending on whether the embryos experience cooler or warmer conditions. Again, however, no intermediate-sex individuals are produced; once the developmental pathway is set, it goes down a series of steps that lead to one kind of individual that makes eggs or one that makes sperm. As Griffiths also says,[5] "The biological definition of sex is not based on an essential quality that every organism is born with, but on two distinct strategies that organisms use to propagate their genes. They are not born with the ability to use these strategies—they acquire that ability as they grow up, a process which produces endless variation between individuals." That variation occurs in shape, size, and behavior, but not in whether an animal is male or female.

So much for sex. What about gender, a somewhat knottier concept? Terms like gender role (as well as sex role) have their roots back to the late nineteenth century,[6] when German psychologists

talked about an analogous concept of sexual temperament, or
Geschlechtstemperament, a word that I am glad we did not import
wholesale into English. Grammar, of course, always used gender
in reference to language, but its use for behavior is more recent. In
the 1950s, noted psychologist John Money popularized the idea in
his work with intersex people. He grappled with how to describe
people's internalized ideas about masculinity and femininity, and
coined the phrase gender role, which by the mid-1960s was used by
both psychoanalytic writers and feminists examining how society's
expectations affected women.

More recently, the word *gender* has sometimes been used syn-
onymously with the word *sex*, for both people and other animals.
Evolutionary biologist David Haig analyzed the titles[7] of over thirty
million academic articles published between 1945 and 2001 for
the use of "gender" or "sex," and found that the use of "gender"
increased sharply since the late 1960s, probably owing to Money's
work, and that the "distinction is now only fitfully observed." He
suggested that gender might be seen as a euphemism for people
trying to avoid mentioning sex.

Does that blurring of sex and gender pose a problem? Perhaps.
An article[8] from 2017 exhorts the reproductive medical literature to
stop referring to the gender of embryos and confine themselves to
sex, in part because both intersex and transgender people do not
find that gender is something that can or should be assigned. In a
somewhat less fraught vein, some fisheries biologists complained[9]
that authors of articles about fish often mentioned gender, with
statements like, "Gender [in largemouth bass] was determined by
visually examining the gonads [testes or ovaries]." They speculated
that the use of gender came about through "a misplaced form of
political correctness" or as a simple substitute to make the writing
sound more interesting, though I am not sure why readers would
be more riveted by a paper about the stability of cod populations
if it included mentions of gender as well as sex. Regardless, the
authors said that the "continued misuse of gender in the work of

fisheries professionals can lead to a lack of clarity, misperceptions, and . . . an erosion of respect for our work."[10]

What about the question of gender in nonhumans? As a question on the website Quora asked:[11] "If gender roles are cultural constructs, how can there be gender roles in animal species?" Along the same lines, another Quora reader wanted to know,[12] "What do feminists think of distinct gender roles in other species, for example, in chickens?" Leaving aside that feminists have probably not gotten together and issued a position paper on gender in chickens, not to mention why chickens seemed like a quintessential example of gender roles, it is a reasonable question to consider. Biologist Jay Schwartz suggested[13] that bonobos and chimpanzees may well exhibit socially determined sex-related behavior, which he considers evidence of gender. For example, in the wild, males usually groom females and vice versa, while in zoos, grooming patterns are based on personal relationships and not just on sex. Other observations from wild chimps show that female youngsters spend more time than males watching their mothers using tools to extract termites from holes, and hence end up learning the technique earlier than males.

Although Schwartz considers this possible evidence of "social determination of at least some, but probably not all, sex roles in chimps,"[14] I am not convinced. Part of the problem is the conflation, as I pointed out previously, of sex and gender. As in the coucals, animals certainly have what we call, for better or worse, sex roles—sets of behavior usually, but not always, associated with males or females. In humans, in addition to sex roles we also have gender, culturally constructed expectations about masculinity and femininity that may vary from place to place or time period to time period. The two are different. In animals, we do not know what the expectations of behavior are, and indeed even asking the question seems a bit pointless, especially when we consider the possibility of gender in animals like fleas or flatworms. It's not that animals never punish group members that behave inappropriately (such as when a subordinate individual challenges a dominant one), rather it's that preexisting, culturally variable norms (like which sex wears which color) are absent. Our ideas about self are complicated enough without trying to impose them onto other species. And Schwartz admits that we would find it difficult, if not impossible, to know if animals have an internal sense of their gender.

My own view is that keeping the terms *sex* and *gender* separate, and reserving gender for humans, is the way forward. This does not mean that sex is entirely innate and gender is entirely environmental, for either humans or animals. Indeed, like all traits, both are influenced by the interaction of genes and the environment. As Griffiths says, how one uses sperm or eggs in reproduction is influenced by the environment, and embryos develop all the characteristics of maleness or femaleness, whether that is a peacock tail, enlarged breasts, or a beard, through a series of steps that require both genetic and environmental input.

Gender, Sex Roles, and Role Models

Even if gender isn't a consideration for animals, sex, and its associated behavior, still is. I noted previously that scientists use the term *sex role* to refer to many different kinds of behavior that are often associated with one sex in animals. A species with gaudy males that display to females and fight with other males, while the female chooses from among several suitors and takes care of the offspring, would be considered to have conventional sex roles; while in other cases the sex role is reversed, such as in Black Coucals, with their large females and dutiful fathers. Our understanding of the basic principles behind these male and female behaviors stems partly from Darwin, who famously noted that males were showy and females coy, attributing these differences to sexual selection, a process in evolution that is similar to natural selection but is based on the ability to attract mates rather than to survive.

Because females produce relatively few eggs over their lifetime, while males can generally replenish sperm rather readily, females often benefit by mating with only the highest quality male, while males often benefit by mating with as many females as possible. These are broad stroke descriptions, however, and they have numerous variations. Females can be competitive, males can be choosy, and the challenge is figuring out what circumstances lead to the evolution of each.

The idea of a sex role, then, suggests to some that this variability is ignored, and this makes people think that there is a normal, or correct, way to do things. In a thought-provoking paper,[15] Swedish biologists Malin Ah-King and Ingrid Ahnesjö point out that talking about sex roles in animals "gives a false impression of biological or evolutionary lawfulness" in animal sexual behavior. And certainly the scientific literature has been rife with the kind of anthropomorphism that makes animal sex life look like a 1950s situation comedy: female birds that mate with multiple males used to be commonly referred to as "cheating," as if they had broken some kind of avian marriage vow. It is easy to project one's own expectations and stereotypes onto animals, whether or not those fit. Ah-King and Ahnesjö also note that in the scientific literature, assignment of sex roles sometimes seem to rely on differences in who takes care of the young, sometimes on who competes more vigorously for mates, and yet other times on which sex reproduces more. Some species have one of these attributes, others have many, and it isn't easy to pigeonhole them all into one category or another.

This inconsistency, and indeed the argument over whether to call something in animals a sex role or not, might seem like it is a matter for academics to sort out, but it colors how we see the nature of maleness and femaleness in other species as well as our own. If people see males as "normally" being competitive and dominant, and females "naturally" being coy or passive, it's an easy step toward accepting the inevitable brutality of the male libido lamented in the *New York Times* that I discussed at the beginning of this book. Similarly, if species with caring fathers or sexually active females are called sex role reversed, such behavior may seem exceptional, something to be dismissed as an aberration. No one wants to be seen as reversed, after all.

At the same time, we don't want to throw up our hands and say that anything goes, and declare that all attempts to come up with a way that male and female animals are likely to behave are politically motivated and have no basis in reality. Across the animal kingdom, males *are* more likely to have flashy displays, females *are* more likely to care for the young, and those patterns arose because males usually benefit more by seeking additional matings and females benefit more by being picky about their mates. This still doesn't lead us to the inevitability of patriarchy, and it doesn't preclude us from rethinking whether talking about conventional or reversed sex roles is a good idea.

Whatever we decide to call them, species in which males care for the young have a lot to teach us about the flexibility of behavior. Take, for example, giant water bugs. These "true" bugs (meaning that they are part of a group of insects with soda straw–like mouthparts, rather than just being something small and icky that people call a bug) live in freshwater ponds and streams in many parts of the world. When females are ready to lay eggs, they place them somewhere they will be safe from harm: the broad back of the male. Males then swim about looking a bit like a beaded evening bag and even guard the tiny bugs when they hatch out, protecting them from fish and other marauding predators. A male can accept eggs from more than one female, mating with each before he allows her to use his back for a nursery, and a female can go on to find another male if she produces more eggs. It all sounds idyllic, except that male back space becomes limiting as females deposit more and more eggs. If females become too numerous, they begin to compete for access to a male, and will even interrupt a male and female while he is fertilizing the eggs. If females are relatively scarce, on the other hand, there is plenty of male real estate to go around, and females are their more customary choosy selves. Similar cases where females become more competitive and less choosy under some circumstances but not others have been reported in some species of fish as well as a number of other insects.

One might assume that when we do see sex role reversal, the typical patterns of sex hormones like testosterone would also be reversed. But in yet another illustration of how neither genes (or hormones) nor the environment control gender, a survey of bird species—unfortunately not including the coucals—showed[16] that testosterone levels were sometimes, but by no means always, lower in males and higher in females of sex role-reversed species than in comparable species with conventional sex roles. In addition, the relationship between hormones and sexual behavior presents something of a chicken-and-egg problem, and not just in birds: an animal's social experience can influence its hormones, which then affect its behavior, which in turn alters hormone levels. This complicated interaction means that we can't just change an animal's behavior by flipping a switch on its hormone levels.

Google and the Male Libido

So if female animals aren't "naturally" always one way and males the other, where does that leave us when it comes to analyzing our own species? Put another way, what difference does it make that male water bugs are caring fathers?

The water bugs matter, I think, not because we should put them on Father's Day cards instead of the ubiquitous golf clubs or neckties (though, come to think of it, that's not a bad idea), but because they remind us that evolution doesn't only have one solution to the problem of propagating genes. This doesn't mean we should use water bugs as role models or insist that all men be primary caregivers of children. There is no point substituting one set of stereotypes for another. It does mean that the biology of sex differences can allow for more than just males who seek to dominate females and females who are only interested in home and hearth.

The water bugs also allow us to question that idea of sex roles—or gender roles, in the case of humans—being innate. Stephen Marche's 2017 *New York Times* op-ed,[17] published in the wake of some of the most egregious #MeToo revelations, asks, "How are we supposed to create an equal world when male mechanisms of desire are inherently brutal?" Marche doesn't invoke genes, or evolution, but the implication is there: masculinity is always threatening a civil society, and we must keep it in check with morality and culture.

But what if that idea about male brutality being inevitable is wrong? Certainly, genes affect male behavior, in humans and in other species, which is what I assume Marche means by inherent. But all behaviors, remember, are the result of a complex interplay between genes and the environment, which makes that inevitability, well, not so immutable. Evolved sex differences exist, but they do not arise in a vacuum unaffected by learning and culture.

This brings us to another commentary written the same year: James Damore's memo[18] to his then colleagues at Google. In brief, Damore questioned the utility of efforts to bring more women into tech fields and hence the company. He maintained that because men and women are psychologically different in ways that affect their likelihood of success at tech-heavy fields, and because those differences come from biology, trying to encourage more women to enter Google was not going to succeed. Women, he claimed, are interested in people, while men are interested in things, a comparison that is not original to him. The resulting uproar ended up with Damore being fired, and I am not going to discuss whether that was justified. But much of his argument centered on the same point made in the Marche's op-ed about the male libido—that certain sex differences are inherent, and you can't do anything about them, so we should stop trying to act as if we can. Neither Damore nor

Marche talked about animals, or for that matter about evolution. But animals, and the way behavior evolves, are important to our understanding of what that inherence means.

Damore says, "On average, men and women biologically differ in many ways."[19] No argument there; it would be bizarre if male and female coucals, and hamsters, and fruit flies differed, but only humans did not. Yet some of the people disagreeing with him seem to claim that while those differences exist in characteristics like height or (duh) the presence of gonads, other, psychological, differences are because of social factors, no biology involved. Here's where I take a different tack. All traits, no matter whether they are behavioral or physical or some of both, need input from genes and input from the environment. Some have more input from one source than the other. But trying to argue that any particular difference between the sexes is completely due to "biology" or "genes" is as pointless as arguing that such a difference is completely learned.

At the same time, "influenced by biology," as I have been arguing throughout this book, is far different from dictated by it. Almost every time we try to pin down one of those supposedly well-established and genetically determined differences, it turns out to be socially influenced. One of the classic sex differences—that boys and men are better at math—turns out to be highly variable, depending on what people are told before taking a test (girls who are reminded of the stereotype do worse than those who are not), the country in which the tests are administered (Japanese girls, for example, do better than American boys), and so forth. Damore also suggests that hormones account for "a lot of differences in career choice," but hormones do not even explain why male shorebirds and not females care for their young. Animals show us that the social and genetic intertwine in us all.

When we think about how boys and girls grow up, and how their behavior differs, one of the first things people often point to is the tendency for children to play with gender-typical toys. This subject has attracted a lot of research, some of which extends to monkeys rather than humans. The reasoning there, it seems, is that if monkeys, which are presumably free from cultural influences, also choose gender-typical toys, that points to an innate tendency to do so.

With children, as you might expect, the studies showed[20] that boys preferred, or at least chose, boy-targeted toys such as trucks and girls chose plush toys and dolls. The differences between the two get larger as older children are tested, and the preferences are consistent across many studies. But toy preferences also depended on how they were measured; the biggest difference between boys and girls occurs if the children are asked to choose only one toy from among several in front of the investigator doing the experiment, rather than when the children are left to play freely among themselves. One problem with interpreting these results is that toys are defined differently depending on who is doing the experiment, with stuffed animals sometimes being classified as girl-related and sometimes as neutral, while blocks are sometimes said to be boy-focused toys and sometimes neutral.

The picture gets even murkier when one looks at the preferences of monkeys. The researchers presented captive rhesus macaques with a variety of toys in their enclosure and then observed them as they played.[21] The study's authors concluded that since the males played more with the wheeled toys, while females liked what the researchers considered more feminine toys, such as a red cooking pot or a plush animal, such gender-typical preferences must be innate. But again, it isn't clear how a female monkey would know that a pot is used for cooking, much less that females are expected to take on this chore. One could make a similar argument about male monkeys playing with cars having little or no awareness of driving. Additionally, since some studies classified the stuffed toys as female-typical and sometimes as neutral, interpretation of preference is difficult.

How might we come up with a definitive answer? In theory, we could do an experiment to see whether, all else being equal, men and women would end up with a difference in toy preference, math ability, the desire to crochet, or anything else. The key phrases here are "in theory" and "all else being equal." What you would need to do is raise boys and girls in a society that treated them identically—and I mean identically—and then see how they behave. Alternatively, we could take genetically identical individuals, divide them into two groups, and subject each group to "boy-favoring" or "girl-favoring" environments, whatever those might be.

Either way, this would be impossible. Study after study shows that people impose varied expectations on males and females from *before* a child is born. An intriguing 2018 study[22] led by Ganna Pogrebna from the UK showed that upon being told they are pregnant with a girl, both mothers and fathers became more risk-averse, as measured by a standard psychological questionnaire. Mothers also interpret videos of crawling infants differently if they are told that the babies are boys or girls, attributing more strength to the former. Again, this is not to say that genetic, or biological, differences are unimportant. Those differences, however, are so entangled with social effects that laying the differential employment of men and women at Google at their feet is ill-founded, along with other absolute distinctions between the sexes. We forget that specious biological absolutes have been common throughout history. A century ago it was claimed that women who were educated would lose the ability to reproduce because their brains couldn't withstand intellectual effort during menstruation. Any genetic explanation for the paucity of women in tech deserves the same scrutiny—and scorn—we would now bring to such a suggestion. And as the water bugs remind us, depending on the circumstances, what we think of as typical male or female behavior can change. If that's true for water bugs, we at least ought to allow for the possibility of the same flexibility in us.

Of Mice and Women

Sex differences in animals, and the evolution of behavior, appear in another way that is relevant to humans, namely their use in scientific, particularly biomedical, research. Many scholars have observed that for decades if not centuries, males—not just in humans, but in all animals we study—tended to take priority, so that even if scientists were examining, say, kidney function in rats, or response to stress in monkeys, they would almost exclusively use male rats or monkeys, and ignore females. Even the description of female anatomy can get short shrift; a 2020 article[23] points out that the Terminologica Anatomica, which as its name suggests lists all the anatomical words in use, has fewer terms for female reproductive parts than for those of males. Yet women and men often suffer different effects from the same diseases, and may respond to drugs and other treatments differently as well.

In response to such concerns, the National Institutes of Health (NIH), which sponsors much of the medical research in the United States, developed a policy[24] that instructs scientists to "consider sex as a biological variable," and include both sexes, human or otherwise, in all of their experiments. The policy was first set forth in 2001, with modifications since then, and has changed much of how diseases and disorders are studied. Many scientists have applauded this goal as a step toward gender equity in medicine, and I think there is much to support about it.

At the same time, the policy may have some unforeseen consequences. Believing that males are just like females is wrong, but so is believing that rats, whether male or female, are like any other animal, or that they are exactly like humans. Finding a sex difference in, say, a response to a drug in a mouse doesn't guarantee that the same response will hold in a rat, or a squirrel, or a monkey, much less a human. Each species evolved in its own way, and while many aspects of the brain and body are similar in mammals, or vertebrates, assuming that sex differences always apply isn't a good idea. A thoughtful paper[25] by Lise Eliot from the Chicago Medical School and Sarah Richardson, a historian of science at Harvard University, points out that even though many mental and brain disorders differ in men and women, adding female mice or rats to studies that would previously have used only males won't be the panacea some had hoped.

One complication is that the degree of difference between males and females, and even the characteristic male and female behaviors, varies enormously among species, as I've already discussed. I would never suggest using our friends the Black Coucals for examining human mental disorders, but if one did, and found a difference between the males and female birds, that difference would likely not be the same as if we had used White-browed Coucals, given their very different mating behavior. Even in rodents, differences abound. For example, many brain and hormonal differences have been found between prairie voles, which pair up and raise their offspring together, and the nearly identical-appearing meadow voles. Just using one of those species would have led to some erroneous over-generalization about what males and females of any species are like. How, then, should we choose a good animal in which to test ideas about Alzheimer's disease, which affects women more than men, even after accounting for differences in longevity? Animals are not just crude versions of people—each species evolved in a particular place with particular circumstances.

When we find sex differences in our test animals, it can also be hard to know how to interpret them. A common test of how mice respond to something that has caused fear—perhaps a loud sound or unpleasantly hot surface—is to see how they react when placed in a new environment. It turns out[26] that female mice are more likely than males to freeze up and not move, a difference reflected in the brain activity of each sex. What does that mean? It could be that female mice can't tell the difference between an old stressor and a new one, or it could mean that for female mice in nature, extra caution in stressful environments is warranted and hence the response is adaptive. Male and female rats also respond differently if they need to forage for food in an environment with a risk of electrical shocks: males increase how much food they collect at one time, but females do not.

None of this suggests that we should go back to the old days of assuming that everything is like a male, or that the NIH has created more problems than it has solved when it instructed scientists to include both sexes in their experiments. But incorporating sex, and thinking about sex differences, in studies of disease as well as of the normal functions of our tissues and organs, is likely to be more complicated than it might at first appear. This might seem to go without saying, but Eliot and Richardson point[27] to the eagerness of pharmaceutical companies to develop "pink" and "blue" therapies. They caution that "as manufacturers have learned they can augment their sales of everything from cell phones to soccer balls in gender-coded colors," drug companies may follow suit, with little justification for fixed differences between men and women.

Avoiding the Essentials

I have already noted that supposedly characteristic male and female behaviors in animals can change depending on the environment, and that females are not always one way and males another. This doesn't mean that sex differences do not exist, or that sex itself is endlessly pliable. But it does mean that using inherent or hardwired differences alone as an explanation misses the point. The hardwired analogy usually means something about our brains, and the brain has become something of a refuge of last resort for arguing that sex differences are unchangeable. But this century has brought the recognition that the human brain is, as British neuroscientist Gina Rippon says,[28] "exquisitely plastic, mouldable by experience throughout life." Despite this discovery, Rippon also laments that current "gender stereotypes about what females and males can and can't do seem to be more rigid and prescriptive than ever before."

Part of the disconnect comes from the media, which often emphasizes studies that show differences between male and female brains, particularly those that seem to reinforce gender stereo-

types about, say, women's inability to read maps or men's unwill-
ingness to share their feelings. Brain imaging studies can give the
impression that sophisticated technology allows us to read people's
thoughts as they have them and, as a result, see how brain function differs among groups, but that kind of futuristic mind-reading
isn't the way brain imaging works. What is more, researchers' own
biases may creep into the interpretation of their results. In one
case, sex differences in brain connectivity were suggested to reflect
men's greater ability at spatial orientation and women's facility at
language, but neither trait was actually measured in the experiment. This study[29] used functional brain imaging to examine men
and women while they were doing a semantically challenging task,
as well as a more mundane task of judging font size. Differences
were indeed observed between the sexes, but men had "greater
tendency for regions to form strongly connected communities . . .
during language processing" and women had "a better balance of
specialized and integrated processing to support efficient information transfer during the language task." Whether that makes either
sex more verbally adept, or even whether the differences found in
the study reflect genetic differences rather than ones induced by
experience, is unclear, though it would be easy to leap to an interpretation one way or another.

And speaking of language, differences in the ways that men and
women express themselves is one of the last bastions of claims for
instinctive differences between the sexes. If, as I've already mentioned, the human exceptionalism of language is a hill that people
are willing to die on, a hummock on that hill is the existence of fixed
sex differences in linguistic ability. As Deborah Cameron, a linguist
at the University of Oxford, points out,[30] people often think that two
such differences are well established: first, that women are innately
better at verbal tasks, and second, that the sexes (again innately)
differ in how they communicate, with men being more competitive
and women more cooperative, or put another way, more inclined to
gossip. Yet neither of these claims holds up well under scrutiny.

Evidence for greater verbal ability in girls or women often comes from responses to standardized tests, or to observations of speech by psychologists and linguists. The tests use such metrics as whether speakers use grammatically correct forms or more informal ones, and indeed, women do tend to speak more "correctly." But as Cameron notes,[31] who is to say that using "isn't" rather than "ain't" means one has greater verbal ability? This tendency is also not universal, with differences depending on the time and place where the tests are administered. What is more, those distinctions of who uses grammatically correct language and who doesn't (or should I say *does not*) also apply to many other categories, such as social class and at times ethnicity, but no one is falling over themselves to argue that those categories are genetically based.

Observations of people speaking in real life, rather than in a test situation, yield similarly mixed results; many studies have shown that women do not talk more than men, for instance. Women tend to use more polite forms of language, at least in some cultures, and that is often associated with verbal ability, but that may arise either because men, as the more dominant group, don't need to bother with niceties of phrase to get what they want, or because women's language is more closely policed. Moreover, although girls tend to outperform boys in educational achievement, that difference may be explained by a number of things other than gender. Boys, for instance, may feel that doing well in school is at odds with being seen as appropriately masculine by their peers.

Let me emphasize, yet again, that finding social influences on the way men and women use language does not rule out influence from the genes. I almost wrote "underlying" influence, but then omitted that word because I think it contributes to the idea that genes, or an evolved predisposition, underlie characteristics, particularly sex differences, like a layer of immovable rock below the shifting seas of social influence. Damore's memo, and many other writings on modern inequities between the sexes, supposes that such inequalities are inescapable. As Cameron says,[32] people can assume that now "we've got rid of the gross injustices of the past . . . whatever inequalities remain between men and women are never going to be eradicated, because they are the consequences of evolved differences that are part of our nature as a species." But all evolved differences, whether in humans, mice, or coucals, represent both genes and the environment, inextricably entangled. It also goes without saying, or should, that the idea of our having rid ourselves of gross injustices is overly optimistic.

I've written a book[33] about the way that our misconceptions about our evolutionary past lead us to make unwarranted assumptions about our present behavior. It is easy to imagine that back in our hunter-gatherer past, we evolved fixed gender roles that, regrettable though these might be under current enlightened conditions, we are now stuck with. In fact, however, though men and women differ both now and in the past, constructing a simple story—what I call a paleofantasy—about the relevance of those differences can lead us astray.

Take, for example, women's behavior during pregnancy. Many guides for pregnant women, including many online groups and advice boards, talk about "nesting" behavior, in which pregnant women supposedly have an irresistible urge to clean house, rearrange their living rooms, and restock their pantries. This activity is said to be instinctive, and one book, quoted in a 2020 paper, affirmed, "The need to nest can be as real and as powerful an instinct for some humans as it is for our feathered and four-legged friends."[34] The behavior is often cited as crazy or irrational. Nesting is also claimed to be both hormonally driven and evolutionarily adaptive, though it's hard to imagine that it would have taken particularly long for early woman to organize the rocks in a cave during prehistoric times.

I find a number of things about this notion odd, not least of which is the term *nesting*, and the restriction to birds and mammals. Insects and fish, I feel compelled to point out, produce some amazing nests, which they then often use to care for their babies.

Perhaps thinking of themselves as a wasp building a burrow in the sand and stuffing paralyzed spiders down it to feed one's larvae just doesn't appeal to some people. And anyway, nests aren't cleaned-up versions of where one lives, they are entirely new structures. One might better imagine pregnant women who are overcome with the urge to answer advertisements for construction worker jobs.

All of that aside, one might question whether nesting behaviors during pregnancy is indeed, as that 2020 paper,[35] by Arianne Shahvisi, is subtitled, a "Biological Instinct, or Another Way of Gendering Housework?" Shahvisi suggests that pregnant women may simply be bored, or feel compelled to clean preemptively because they know they will have less time when the baby is born. I can't even imagine how we would test the idea. The point is that the presence of the behavior, or its commonness, or even feeling it as a compulsion, doesn't mean it isn't socially constructed. Rather than searching for hormonal or other biological explanations for pregnancy-related nesting, it might be worth seeing how, as Shahvisi says, "Gender stereotypes lead to misuses of biology and consign women to particular social roles."

The essentialist view—that gender, not sex, has a fixed, inherent nature—itself affects the way we learn, and think, about genetics itself. An intriguing study[36] of eighth to tenth grade students in the United States first asked them to read a genetics text that either explained plant sex differences, sex differences in humans, or provided a discussion of how genes did not dictate behavior or ability. The students who read either of the first two texts were more likely to believe that a person's ability in science is innate as compared to the students who were offered the discussion. These outcomes held particularly true for girls. In addition, a belief that the ability to succeed in science is inherent predicted a lower level of interest in scientific careers for the girls, though not for the boys. These results point to a self-fulfilling prophecy; if people believe they will not succeed in something because of their gender, and that attitude makes them less likely to do well in that area, then we can point to the lack of successful members of that gender as evidence of a lack of ability, even when no such lack exists.

Finally, back to the animals. On the Quora question about animal sex and gender, one of the comments was a bit plaintive, wondering why animals, unlike humans, are free from culturally imposed gender roles and can "live the way they want."[37] Whether all those animals actually do live the way they want is questionable, and since the area of free will is fraught enough for humans, it seems even more unanswerable for the animal world. Nonetheless, whether we look at lions, dogs, water bugs, or coucals, animals show us that the array of behaviors associated with being male or female goes far beyond what humans might think.

11

Protect and Defend:

Behavior and Disease

After the experience of the last few years, it would be hard to overestimate the impact of infectious disease. The COVID-19 pandemic changed virtually everything about our lives, from how we worked and played to the overuse of words like "unprecedented" and the phrase "we are all in this together." Plagues have always been with us, of course, and the most recent pandemic will not be the last. Humans have been dealing with disease since before we became modern *Homo sapiens*, and we have and use many tools to protect ourselves. Our immune systems evolved to fight disease, and the triumph of vaccine development is nothing short of breathtaking. But for millennia, the first thing we have been doing to combat illness is to change our behavior. Indeed, for the first months of even this most recent pandemic, humble changes in behavior were all we had: washing our hands, keeping our distance, putting on a mask.

Although animals do not use hand sanitizer or wear masks, they too are subject to diseases, and those diseases can alter their behavior in two major ways, one that helps the infected individual and one that harms it. The helpful kind of behavior change is similar to what humans do when they are sick—fumigate their homes, perform triage on wounded community members, and eat special foods to help them heal. The harmful kind is instigated not by the sick individual, but by the organism that has infected it. In this chapter I discuss both, and how behavior can help animals and humans cope with disease.

First, a few words about terminology. I am confining myself to infectious diseases, which includes more serious threats like COVID-19 and bubonic plague as well as relatively minor ones like the common cold and ringworm. Infectious diseases are those that are caused by other living things, such as bacteria. That distinguishes them from illnesses like hypertension or diabetes, in which the body has a defect that was not caused by another creature. The distinction is important from an evolutionary perspective because infectious diseases are a threat in a way that the others are not: they can evolve back at you. Diabetes doesn't evolve to circumvent insulin injections, but the bacteria that causes plague or tuberculosis will evolve to become more or less deadly depending on the circumstances. Any microbe that happens to have the ability to resist a particular defense will have an advantage and will out-reproduce its less resistant kin. That in turn imposes selection on a host to evolve an even more effective response. This is, of course, natural selection at its finest. It is also why we are struggling with antibiotic resistance. The back-and-forth has been described as an evolutionary arms race, in which neither party will ever be able to declare victory. Because selection for behaviors that counter the pathogen's effects is ongoing, and new defenses are continually required, anti-disease behaviors are one of the best places to look for the way that behavior evolves.

Another point about terminology is really a way to broaden our perspective about the cause of disease. From the point of view of an animal or person that gets infected, it doesn't much matter whether the infection is caused by viruses, bacteria, or tapeworms. Although the latter are usually considered a parasite, and not necessarily a disease, the distinction does not mean much. The treatment differs, of course, between viruses and tapeworms, but it also differs among various kinds of microbial infections. Worms can be welcomed into the fold, so to speak, along with their smaller disease-causing cousins.

Whose Side Are You on, Anyway?

As I noted at the open of this chapter, behavior that evolves in response to disease can actually harm the infected individual. Being sick is generally bad for the one infected and good for the organism doing the infecting, of course, so when a diseased animal does something that makes itself worse, an explanation is needed.

That explanation lies in what is termed parasite manipulation of host behavior. Ants, for example, when infected by a particular fungus, climb to the top of a blade of grass or a leaf and clamp onto it. Although uninfected ants would never dream of such a mountaineering feat, scaling the grass blade fosters the growth of the fungus and allows it to release its spores at the best possible place. Fish infected with worms will swim close to the surface of the water with their shiny white bellies exposed, again making them easy prey for birds, the final host for the worm. And a lovely insect called a jewel wasp injects venom into a cockroach, rendering her prey docile enough to lead by its antenna to her nest, where she places her larvae on it. The young wasps then have a catatonic and perfectly preserved food source to munch while they grow. In this case, the wasp is more akin to the flies that parasitize my Hawaiian crickets than to a disease, but the manipulation is just the same.

Perhaps the best-known example of parasite manipulation is when mice lose their fear of cats. This happens when the mice are infected with a one-celled parasite called *Toxoplasma*. The parasite occurs in many kinds of mammals all over the world, including humans, but it reproduces only in members of the cat family. In people, toxoplasmosis, the disease that results from an infection, can cause miscarriages, which is why pregnant women are urged to avoid cleaning the litter box.

Cats acquire *Toxoplasma* by eating infected mice or other prey. As you might imagine, anything the parasite can do to increase the likelihood of its rodent host being eaten by a cat would evolve under natural selection. How to make a mouse more vulnerable to being eaten by a cat? Simple: make the mouse less afraid of cats. And indeed, that is exactly what the parasite seems to do. Experiments dating back to the 1990s show[1] that *Toxoplasma*-infected mice and other rodents are actually attracted to areas scented with cat urine, whereas their uninfected counterparts would retreat from the odor. The fearlessness would, of course, make the infected mice more vulnerable; hence, the parasite manipulates its rodent host and thereby increases its chances of getting into the final host where it can reproduce.

The idea that parasites manipulate host behavior to achieve their own ends has become widely accepted, and stories about such behavior are often in the media under headlines that contain words like "zombie," "hijack," and "fatal attraction." Like many such stories, however, the truth is a bit more complicated.

Two recent studies have questioned the conventional wisdom about the way that manipulation happens. The first was a study[2] examining the brains, genes, and behavior of mice that were infected with *Toxoplasma*. As in previous work, the experiment allowed mice to go into areas scented with cat odors—in this case bobcat—as well as areas scented with odors of other animals, including foxes and guinea pigs. Infected mice did explore the cat-scented places more than the uninfected mice did, but, somewhat to the researchers' surprise, the infected mice also went into the other areas more often than their healthier counterparts, even though guinea pigs are not going to get *Toxoplasma* from a mouse under any circumstances. The investigators found that infected mice had generally lower anxiety levels and were more exploratory overall, which they said explained the altered behavior and meant that the parasite wasn't responsible for the change. Coverage of the work suggested that the old story had been debunked.

I am not so sure. After all, infected mice are still more likely to be eaten. Just because they might waste their time getting up close and personal with a guinea pig, that doesn't mean their decreased anxiety levels won't get them eaten by a cat at some point. Or, as parasitologist Laura Knoll at the University of Wisconsin-Madison put it in a commentary[3] on the work, *Toxoplasma* "clearly manipulates the crap out of the host." A generalized loss of anxiety due to changes in the brain is a much more plausible mechanism than one that requires the parasite changing the mouse in such a way that it is less fearful of a cat, but only a cat. Once again, complex behaviors can evolve using relatively simple mechanisms, and they also do not require any diabolical intent on the part of the one-celled *Toxoplasma*.

That diabolical intent, along with all the sensationalized language about zombies and mind control, was the subject of the other recent critique[4] of parasite manipulation. Jeff Doherty, a scientist at the University of Otago in New Zealand, thinks that scientists like me as well as the general public are too quick to embrace what he terms "inherently vague and misleading" metaphors about parasites. Whether it is due to the influence of sci-fi and horror films like *Alien* or simply out of a desire for a good story, he says we are succumbing to quick and easy explanations that are often inaccurate.

Is he just being a spoilsport, even a killjoy, in an age when people may not always want to read about science without a bit of added punch? Maybe. Many of my colleagues think it's fine to personify animals, plants, and even one-celled organisms as a way of encouraging the public to embrace nature and find otherwise peculiar or—let's face it—dull creatures interesting. I shouldn't be casting the first stone here, either, having done my share of using human terms to describe animals. At the same time, such personification can lead to misconceptions about the way behavior evolves. If we emphasize the idea of a puppet master when we talk about a single-celled parasite, it's hard not to visualize an animal behaving the exact same way a person would.

It seems to me that it's even more marvelous to think that such behavioral manipulation can happen without intent. Indeed, that was demonstrated in a recent study[5] of the ants that crawl up onto a leaf and die when infested with a fungus. A group of researchers from Pennsylvania State University used highly sophisticated imaging techniques and computer learning methods to visualize the fungus inside the body of an infected ant. Somewhat to their surprise, the fungal cells never even enter the ant's brain. Instead, they surround its muscles and manufacture compounds that directly influence the ant's movements. No mind control is necessary if the mind, and the brain, aren't involved. And it means we don't need to wonder how the fungus manages to act like the world's best air traffic controller.

Social Undistancing

We humans think that we are above such manipulation, and probably assume that in any battle of wills between ourselves and a parasite outside the movies of Hollywood, we would be able to emerge triumphant. But the fact that these parasites don't need any special mental abilities to alter the behavior of an ant or cockroach, and that the hosts' higher cognitive powers aren't involved either, means that manipulating humans is perfectly within the realm of possibility. As with all behaviors, individuals who act in a certain way just need to be better at surviving and reproducing than the ones who do not act that way, and they need some way of passing on their abilities.

So, remembering the example of a fungus that does better by making its ant host climb up a grass stem, what human behavior might benefit a pathogen that infects humans? Pathogens that are spread by contact between infected individuals do better by increasing the likelihood of that contact, and in highly social creatures like humans, making humans even more social should do the trick. You cannot, of course, measure the predilection for party-going in people after injecting them with a disease, but biologists from Colorado State

University and Binghamton University hit on a clever way to test the idea[6] by taking advantage of the next best thing: vaccinations.

Vaccines stimulate the immune system, which at least on a short time scale imitates the effect of a real disease. So the scientists asked if getting the flu vaccine, as many people do when the season is upon us, would alter the social behavior of those who received it.[7] They noted that if such a response existed, it could favor either the virus—because sick people would spread a disease more if they contacted more potential hosts—or the infected individual, perhaps because they would solicit more care from their companions. They used people getting flu shots at a campus clinic, telling them they were participating in a study about illness and social behavior but giving no particulars. At three time periods, the subjects were asked to reconstruct their activities during the previous two days: before the shot, forty-eight hours afterward, and four weeks later. Even after controlling for variables like the day of the week the shot was administered, the occurrence of holidays, other special events, and more, the results were striking: in the forty-eight hours after their immunization, people were much more social than they had been beforehand, although none of the subjects self-reported a sense that their behavior had changed. The boost disappeared at the four-week measure.

Whether this was the doing of the virus or the host can't be disentangled from the study, but it certainly suggests that we can consider ways that the evolution of behavior affects how we treat disease and what we can expect from various public health interventions. The paper was published before the COVID-19 pandemic, and it is tempting to wonder whether social distancing is psychologically made even harder by an infection. Along these lines, a 2020 paper[8] by scientist Patricia Lopes musing on the effects of the pandemic noted, "By adopting social distancing as part of our battle against a novel infectious disease, we are fighting against some of what it means to be human: to live socially."

Noting that viruses can evolve and potentially change their hosts' behavior, even when those hosts are people, shouldn't come as a surprise. Since virtually all animals have anti-disease behavior, the virus needs to counteract those behaviors and make the host behave so that it spreads the virus. In early 2021, as Europe was experiencing a resurgence of COVID-19, Mayor Francesco Vassallo of Bollate, Italy, a town near Milan, lamented,[9] "This is the demonstration that the virus has a sort of intelligence, even if it is a single-cell organism. We can put up all the barriers in the world and imagine that they work, but in the end, it adapts and penetrates them."

We all sympathize with his despair. The virus doesn't need intelligence to overcome at least some of our barriers. Still, we shouldn't be overly pessimistic. The truth is more hopeful; it may not be upbeat exactly, but it's not the doom and gloom Vassallo predicts. A virus can't understand evolution. But the good news is that we can.

A Bitter Pith to Swallow

Parasites and disease do not, of course, always have the upper hand. We do many things to either stave off or treat disease, and some of those things involve changes in behavior. Even early humans used medicine and treated injuries such as fractures, so it seems plausible that our closest relatives, at least, might do the same. And indeed, in the 1980s and 1990s, several researchers, including Richard Wrangham, noted some peculiar behaviors in the chimpanzees they were studying in Africa.[10] Chimpanzees eat a variety of plant parts, including fruits and leaves, but some individuals were selecting young shoots of a plant called *Vernonia amygdalina*, which grows in many parts of tropical Africa, and stripping the stems of their bark. The chimpanzees then chewed on the bitter pith and juice, consuming sections anywhere from two inches long to a foot or more. The individuals that did this often seemed to be sick, exhibiting diarrhea, previous weight loss, and a lack of energy. Wrangham and others, most notably anthropologist Michael Huffman, examined the parasitic worm eggs in the feces of the animals, and found that the use of the plant was associated with a drop in their parasite egg number. Interestingly, the chimpanzees avoid the leaves and bark of the plant, which contain extremely high levels of the chemical thought to act as an anti-worming medication, levels that would be toxic if consumed in any quantity. Chewing bitter pith has now been seen on many occasions, with young chimps occasionally tasting the plant material their mothers had chewed.

A similar behavior with a similarly descriptive name, leaf-swallowing, has also been seen in chimpanzees.[11] Here, the animals take a leaf from an *Aspilia* plant, fold it up, and swallow it whole, without chewing, so that the entire undigested leaf is found in the chimp's droppings. An individual chimpanzee can swallow up to fifty-six of the leaves at a time. Again, the leaves are not eaten as food, and they are swallowed by animals that appear to have gastrointestinal symptoms associated with worms. *Aspilia* leaves are quite hairy, and scientists believe that the hairs help to scrape the worms from the gut, allowing them to be expelled.

More recently, similar behaviors involving leaf swallowing were observed in Asian gibbons,[12] another great ape species, and both leaf-swallowing and bitter-pith chewing have been seen in bonobos and gorillas too. Other apes, including orangutans,[13] as well as monkeys in South America, have been seen rubbing fruit or chewed leaves on themselves, which is interpreted as a way to rid their bodies of external parasites like ticks or fleas, or perhaps repel mosquitoes (cats may perform a similar activity, which I'll discuss later).

How do the apes know what plants to choose, or when to administer them? Huffman and others see these behaviors as having evolved when our physiological means of avoiding disease, such as our immune system, is insufficient, as a kind of medical backup plan, and something that requires a distinct ability and probably a sophisticated level of cognition, given its occurrence in apes. From that point of view, self-medication in animals thus becomes a singular, even extraordinary, phenomenon. But it seems to me that separating such behavior from physiology, and assuming it requires some extra-special powers, causes the same problem I've noted all along: behavior and physical characteristics both evolve in the same way, and evolution doesn't have any reason to use one set of mechanisms first and then try another. Animals have many ways of dealing with infection: some are on a cellular level, some are with tissues and organs, yet others are with behavior. This does not mean the behavior isn't fascinating, or that investigating its origin isn't worthwhile. But it might mean that it's worth taking a step back to see where else in the animal kingdom such self-medication occurs.

Food, Medicine, Fumigation, and Goats

Since the discovery of leaf-swallowing and bitter-pith chewing, many other species, many of them non-primates, have been found to use substances that help kill or control parasites. They do so, mostly, by changing their diets. This has required us to do something of a retrenching in our thinking about what qualifies as self-medication, and to broaden the criteria of self-medication, since the behavior appears to be deep-seated in animal evolution. If only chimpanzees and people self-medicated, it would suggest that such behavior is highly specialized and sophisticated, possibly requiring advanced cognitive ability. But if, as it turns out, the behavior is widespread, and arose many times in evolution, then we have to look more deeply into the behavior's origin.

That broader perspective means we need to define self-medication more precisely. Scientists have more or less agreed on four criteria.[14] First, the animal has to ingest or apply something it ordinarily wouldn't, whether that is more of a usually shunned food item or, as with the chimpanzees, a novel plant. Second, the behavior has to be prompted by the animal becoming ill, something that can be extremely hard to demonstrate without doing a careful experiment, and which means that some instances of supposed self-medication by elephants, tigers, and dogs are merely anecdotal for the present. Third, the behavior has to benefit either the animal itself or its kin. Finally, self-medication has to have some cost; perhaps it involves eating a substance that is also toxic, even though it cures the animal of a disease. If this last criterion is not met, then one would expect all individuals, sick or not, to consume or apply the medicine, since nothing would be lost by doing so. The use of medicine by animals, as in humans, involves balancing the benefit of treatment against the risk of the medication itself.

Instead of selecting a plant that is not ordinarily considered edible, some animals will eat more of one plant species than another, or will change their environment to make it less hospitable to disease. Some of the best evidence for self-medication in animals other than primates[15] comes from what might seem like an unlikely source: goats and their fellow cud-chewing relatives, including sheep. Goats, of course, are known for supposedly eating anything, from tin cans to laundry off the line, but they turn out to be remarkably sensitive foragers. If free-ranging goats are infected with roundworms, they will increase their consumption of a shrub containing a chemical that fights the worms. Interestingly, only some breeds of goats will refine their diets in this way, which suggests an interaction between the genes of a particular breed and its likelihood for becoming parasitized. Goats are particularly sensitive to tannins in their feed. Tannins are a group of plant chemicals that give tea, red wine, and a host of other foods their astringent feeling when you eat or drink them, and are used, as the name suggests, in tanning leather. Their bitter taste helps the plant by deterring herbivores, at least some of the time, but tannins can also act to help rid the herbivore's body of worms (though this is not a suggestion to use tea, much less wine, as a human anti-worming drug).

It is much easier to do controlled experiments on sheep and goats than on apes, and a number of studies[16] have shown that the animals adjust their diets depending on the nutritional content of the food as well as on their health. Lambs that had been infected with intestinal worms, for instance, ate more of a food containing tannins than their unparasitized counterparts, but the difference in diets faded after the first twelve days of the study, which corresponded to the time when the worm levels decreased. Ruminants seem to be able to detect which foods are better for them, also; if sheep are given an internal infusion of highly nutritious liquid close to the time they are offered an otherwise low-nutrient feed, they will eat more of the food they otherwise would have shunned. Whether this means that the animals left to their own devices learn to eat medicinal foods because they associate them with feeling better afterward is unclear, but that would be a potential pathway for the acquisition of self-medication. One possibility is that animals that are sick become more open to eating new foods more generally, something that again has been demonstrated in an experiment with lambs.

Food, in other words, is their medicine, in keeping with an adage thought to have arisen with Hippocrates. Rather than consuming a new plant, animals simply eat a new combination of foods. The ruminants also illustrate the adage that the dose makes the poison. In virtually all cases, the animals do not switch their diets completely over to the tannin-containing plants. Instead, they just recalibrate how much of a given plant they consume. Sheep will even eat polyethylene glycol, a chemical used in many human medications, when they are also consuming foods with a high tannin content— the polyethylene glycol neutralizes some of the less desirable effects of the tannins.[17] It isn't always clear how self-medication develops in an individual. Mother sheep seem to teach their lambs about which plants are best to eat, but the experiments that would be necessary to demonstrate the exact contributions of genetic predilections and early life experience to the behavior have not been done.

Though the animals show a remarkable kind of "nutritional wisdom," as some of the scientists studying them put it, these results do not mean that we could all cure our own diseases if left to choose the right foods, or that tannins should become the new superfood, even for sheep and goats. It does suggest, however, that by allowing grazing animals to consume a wider variety of plants, we might be able to reduce the number of drugs they are now given to control parasites and diseases, drugs that can contribute to antibiotic resistance and pollution.

And speaking of pollution, or at least of human-produced substances that seem to litter our environment, an unusual form of preventing disease has been discovered in urban birds. In addition to being snug places to raise the chicks, many birds' nests are plagued by lice, fleas, mites, and other parasites. These pests are more than a nuisance, since they suck blood from the vulnerable featherless young birds and can lead to slower growth or even death. Such pests are a particular problem in species that use their nests over and over, either within a season or even from one year to the next.

The birds can't physically remove the intruders, but some species do the next best thing: they place aromatic leaves inside the twigs, grass, or other materials used to build the nest. The plants act as a natural fumigant, reducing the number of fleas and other external parasites. In an interesting twist[18] to the story, Constantino Macías Garcia at the National Autonomous University of Mexico and his colleagues have documented House Finches, common birds that have extended their range to most parts of North America, weaving fibers from cigarette butts into their nests. The butts contain nicotine, which is often used as an insecticide, and the scientists wanted to see just how the birds used the novel material and if it was effective. They selected thirty-two nests, and a day after the eggs hatched, took out the lining the parents had already added, replacing it with felt, to ensure that any parasites already present were gone. Then they added live ticks to ten of the nests, dead ticks to another ten, and left the remaining nests parasite-free.

The parent finches (who one can imagine questioning their own sanity: "I could swear this isn't the way the nest looked before") were much more likely to add fibers from cigarettes to the nest if it had ticks, and they added much more of the nicotine-containing material to the nests with live ticks versus dead ticks. The amount of fibers added was also greater if the nest had originally contained more fibers, suggesting that the finches can gauge the parasite levels inside their own nests. The use of tobacco, however, carries a cost, even for birds; both the nestlings and parents from the nicotine-enriched nests showed signs of DNA damage that was not seen in the unexposed birds.

Note that what may look like anti-parasite behavior may not be. Most of us are familiar with cats responding to catnip by licking and chewing it and rubbing their faces in the leaves. It has been demonstrated that some cats seem to enjoy the substance and some are indifferent to it. A group of Japanese researchers wondered if the behavior was recreational or had some other function,[19] and noted that catnip, along with its relative silver vine, can also act as a mosquito repellant. The scientists suggested that the behavior evolved to keep cats from being bitten, which potentially transmits disease, and that the distraction of buzzing mosquitoes might make hunting less effective. The media loved the idea of cats using catnip as a medication, and articles on how to adapt this for human use suddenly sprung up online.

I am not completely ruling out cats using catnip to deter biting insects, but I remain a bit skeptical. No one has demonstrated that cats in the wild use the plants successfully, nor that mosquitoes are a major problem for cats. Unlike dogs, domestic cats do not get mosquito-borne diseases in large enough numbers for concern, and the presence of a substantial fraction of nonresponders (seven out of twenty-five of the cats in the study[20] were unaffected by catnip or silver vine) also argues against it being important for cats. Maybe we just have to accept that cats enjoy a nice recreational substance now and again.

The Not So Very Hungry Caterpillar

Neither ruminants nor songbirds win any awards for their reasoning skills, which makes the notion that warding off illness requires a high level of cognition, an assumption made by some of those studying the practice in apes, questionable. And as with many other complex behaviors, some of the best examples of self-medication come from animals with very tiny brains, namely insects. It would be easy to say either that insects have instinctive self-medication behavior, while primates and other vertebrates must learn to treat their ailments, or that both have some kind of innate wisdom about what's good for their bodies, but that would be wrong on all counts. Like all behaviors, including complex ones like treating one's own diseases, self-medication in primates and insects alike requires both input from genes and environment.

Admittedly, watching a caterpillar determinedly munching its way through a pile of leaves takes some of the romance out of animal self-medication. As with the goats, however, changing their diet is one of the first ways that insects defend themselves against disease. The familiar monarch butterfly is well known for its toxicity to birds that try to eat them; the butterflies get their noxious taste from chemicals in the milkweeds that the caterpillars feed on. Those chemicals, called cardenolides, can help control parasites that cause disease to the caterpillars, but they are also toxic to the caterpillars. Different kinds of milkweed have different levels of cardenolides, and mother butterflies lay their eggs on milkweed with more cardenolides if they are infected with a pathogen than they do if they are healthy. In so doing they confer greater protection on their young. The defense against disease comes at a price, however, with lower survival of the toxin-feeding caterpillars. Presumably this is a price worth paying in circumstances where diseases are rampant. Similar consumption of plants with high levels of pathogen-deterring chemicals has been documented in woolly bear caterpillars as well as bumblebees, which will preferentially consume nectar that contains alkaloids, another set of chemicals that can fight infection.

It's particularly interesting that the monarch butterflies cannot cure themselves of an infection, but can only help their offspring grow up more resistant to attack. Such medication of kin, rather than of oneself, has now been demonstrated in a number of insect species.[21] Perhaps unsurprisingly, social insects such as ants and bees are excellent medicine providers for their colonies, with both kinds of insects collecting plant resins and using them to line their nests. Like a botanical version of the sprays and wipes we use in our homes to kill germs, resins have natural antimicrobial properties. The insects use them enthusiastically, with some ants gathering up to twenty kilograms (forty-four pounds) of the sticky stuff and stashing it in their nest.

Honeybees have long been known to use resins collected from a variety of plants to coat the inside of their hives. A study[22] by my friend and colleague at the University of Minnesota, Marla Spivak, showed that bees infected with chalkbrood, a fungal disease, collected more resin than uninfected control bees, and that the resin protected the hive from further infection. The bees do not eat the resin, making it a somewhat different kind of medical treatment than the leaves consumed by the apes or the dietary changes in goats. Marla advocates[23] for the term *social-medication*, as opposed to *self-medication*, since the practice is not limited to the user of the substance in question. She suggests that because the resins are sticky and difficult to handle, bringing them back to the nests is a costly behavior that nonetheless pays off. That stickiness might also have contributed to an inadvertent problem for modern beekeeping; Marla believes that because the resin makes opening manufactured hives difficult, beekeepers have selected for bees that do not use it as much, which in turn leaves the colonies vulnerable to disease.

My favorite recent example[24] of medical treatment in the insect world comes from a rather obscure ant species in Africa called *Megaponera analis* that makes its living hunting termites and bringing them back to the ant colony for food. Hunting termites is dangerous business, at least for an ant, because the prey defend themselves with their formidable jaws, and an ant may get injured, losing one or more of her legs (as an aside, all the workers that go out on the raids, like all the ants you see at your picnic, are worker females). Individual ants are often considered to be the disposable fast-fashion items of the insect world, cheaply generated and quickly consumed, meaning there is little reason to repair or cherish them.

These ants, though, live in relatively small colonies of several hundred, compared to the hundreds of thousands in other species. This means an ant that is wounded on a raid is often still worth saving, and a lightly injured party will take up a hunched position with the legs drawn up, which makes her easier to carry. Other ants then perform a kind of battlefield triage, determining which ants are worth saving, and leaving behind those that bear fatal wounds, such as having lost more than two legs. After taking the injured ant back to the colony, the other workers will groom and lick the area around the wound, removing the dirt and spreading natural antimicrobials from their salivary glands. Note that dirt removal is a rather specialized activity. Because ants, like all insects, have external hard skeletons that completely cover their bodies, dirt usually poses no threat to them; when their skeleton is punctured, however, disease-carrying bacteria become much more of a threat. The grooming reduced the likelihood of ants dying of their wounds by a whopping 80 percent. The four- and five-legged survivors quickly learn to cope with their disability. Interestingly, about a third of the ants on raiding parties have lost at least one leg. To me, this triage assessment followed by transport and finally treatment is far more sophisticated than a chimpanzee folding up a leaf when it has a stomachache.

Similar medical behavior, though without the field medics, has been observed in many other insect species, including flies, other kinds of ants and bees, and moths. How do these tiny creatures with their limited nervous systems perform such complicated acts? As with the other behaviors I've discussed in this book, there is no need to invoke a mysterious wisdom that comes with listening to the body or paying attention to nature's pharmacy. Instead, it's about applying those same rules of thumb that my crickets use when paying attention to how much song is present as they grow up.

When it comes to self- or social medication, behavior that responds to a particular cue in the environment, say the presence of a fungus in your nest if you are a bee, will evolve so long as the individual doing it survives and reproduces better than individuals that don't. Anti-disease behavior is no different from other behaviors in this regard, and it evolves in stages, just like the spider-mimicking tail of the vipers I discussed in chapter 2. Imagine a primordial caterpillar that ate very slightly more of a food that contained disease-fighting chemicals when it was infected, but not when it was healthy. Why would it do that? It doesn't matter—maybe the leaf smelled different from the rest. What matters is that doing so resulted in marginally better survival, and the predilection for eating it is at least partly heritable. Remember that evolution can move in small increments over nearly unimaginable lengths of time. The caterpillar doesn't need to know what it is doing any more than the snake has to determine exactly how to make that spidery tail appear.

Neanderthal Teeth and Spicy Food

If animals using medicine is so widespread, then when in our evolutionary history did humans adopt the practice? And could we have learned about treating our own illnesses from animals? Several of the researchers working on ape self-medication speculate that humans could well have observed sick animals recovering after eating particular plants, and then emulated them. I am a bit skeptical about this possibility, given the number of hard hours of observation that have been necessary to document the practice in the primates to begin with. It's not likely that early humans had the leisure time to do that much natural history work, but it can't be ruled out. Perhaps more intriguing is the observation[25] that elephant keepers in Thailand use many plants to treat their charges, some of which are also used by elephants in the wild, and 55 percent of them are seen in human medicine as well. Elephants have been part of human culture in Asia for at least four thousand years, and their behaviors may well have been carefully monitored.

Whether it arose from imitating other animals or not, the idea that early humans, or humanlike ancestors, were employing plants to heal their illnesses seems plausible. Even today, more than 80 percent of people in developing countries use plants for virtually all their medical requirements, and the natural pharmacopeia of plants with active compounds is extraordinarily large. Documenting the use of ancestral medicine is difficult, of course, since plant material usually doesn't fossilize well and other evidence is scanty.

An unusual source of information about our evolutionary ancestors and relatives use of medicine comes from a rather humble place: dental calculus. Also called tartar, it is the hardened stuff that accumulates on the teeth and, for modern humans, gets removed by the dental hygienist at our regular cleanings. Early hominins accumulated dental calculus too, and the hardened material traps anything that goes into the mouth. Archaeologists have analyzed[26] DNA from the dental calculus of Neanderthals from several sites in Europe and found traces of chamomile (*Matricaria chamomilla*) and yarrow (*Achillea millefolium*), neither of which are food plants and both of which may be used in herbal medicine to treat a variety of maladies. A penicillin-containing fungus was also found in the teeth of a Neanderthal, giving rise to the speculation that antibiotics might even have been employed in prehistoric times.

Early humans, as well as Neanderthals, may also have helped each other when they were ill or wounded. Skeletons showing signs of recovery from infections or severe injuries including broken bones have been recovered from sites in several parts of the world. Given that we are social creatures, it stands to reason that our ancestors would have helped each other when necessary. This does not mean, again, that we—or any other animal—have a mysterious inner instinct that shows us which plants are poisonous and which can heal us. It is tempting to think that we simply inherited the knowledge found in chimpanzees and other apes and then embellished upon it as our species evolved. Indeed, anthropologist William McGrew suggested[27] that we can use chimpanzees as a blurry model of our earlier selves, saying, "Anything a chimpanzee can do today could also have been done by the Last Common Ancestor six to seven million years ago." He admits that this is a simplifying assumption, but it also falls into the fallacy of the scala naturae: chimpanzees did not stop evolving when they and we split apart from that ancestor, and their self-medication behavior, as well as many other aspects of their lives, could certainly have changed dramatically over such a long stretch of time. We could both use medicine because of convergent evolution rather than a shared inheritance.

Have humans evolved to use food—not specific dishes, but their customary diets—as medicine? One of the most basic ways in which cultures around the world differ is in their cuisines. Scandinavian food is not the same as Italian, and both use different ingredients with different manners of preparation than are commonly found in Thailand or India. Perhaps most notably, cuisines differ in the amount and kind of spices that they use, and it's common knowledge that tropical countries tend to have spicier food. In the late 1990s, behavioral biologist Paul Sherman[28] wondered if this co-occurrence of hot dishes and hot climates might have arisen because the spices help prevent parasites and foodborne illnesses. In other words, cuisines evolved to be spicy where those spices did the most good. If so, then people who used spices where the risk of getting sick from food was high would be more likely to survive and reproduce than people who ate a blander diet. Again, it doesn't matter why people think they use the spices, just that they do so.

It is certainly true that many spices have antibacterial properties, and meat is particularly likely to spoil in places where refrigeration is or historically was hard to come by. Sherman and his colleagues reasoned that meat dishes therefore should be more highly spiced than vegetable ones. They tabulated the use of spices in recipes for both meat- and vegetable-based dishes from thirty-two countries, and then looked for any relationship between climate and the spiciness of the cuisines. As they had predicted,[29] hotter places used more spices, and were particularly likely to use more chilies, garlic, and onion, as well as anise, cinnamon, coriander, cumin, ginger, lemongrass, turmeric, basil, bay leaf, cardamom, celery, cloves, green peppers, mint, nutmeg, saffron, and oregano. Not all of these are spices by a strict definition, but the first ten items, all of which are spices or at least similar in their flavoring use, are especially good at inhibiting bacterial growth.

This seems like conclusive evidence for cultural evolution in humans, with societies gravitating toward foods that also help them avoid illnesses. But more modern work has found a problem with Sherman's analyses, one that is only solved by recent methods for analyzing the kind of data Sherman used. Lindell Bromham, a professor at the Australian National University in Canberra, pointed out[30] that the correlations among cultures make it difficult to draw conclusions just by looking at relationships between a given country, its climate, and its cuisine. It is in part a classic confusion of correlation and causation that can arise in evolutionary studies.

Before delving into the spices, consider a simpler, hypothetical, example. Say you want to see whether birds that wake up before dawn are also more likely to eat worms than birds that rise later in the day. You imagine that selection has acted on each of the species, favoring the predawn behavior because it makes hunting worms more effective. You survey ten bird species and find that, indeed, the three species that sleep late eat other foods, while the seven bird species that are up early mainly consume worms. Conclusion drawn.

The problem, however, is that you didn't control for another variable that led to your results: the evolutionary relationships among your ten species. It turns out that of the seven worm-eating species, five shared a common ancestor relatively recently. They all inherited their predawn wakefulness from that ancestor, and so they really don't comprise five independent ways to test your idea, but only one. You would need to find four more bird species that did not have a recent ancestor to test the theory.

Similar issues arise in any large-scale comparison seeking to explain why traits occur together, and the spices-parasite link is no exception. As Bromham put it,[31] "cuisines in South East Asia are all more similar to each other than any of them is to Scandinavian food, and they are also all in a much warmer climate. Treating all of those cuisines as separate data points will make it look like there is a link between climate and cooking, even if there isn't." The food in Thailand is like the food in Malaysia in part because the populations of both countries learned from each other as their cuisines were developing. Twenty years ago, when Sherman did his analysis, techniques for statistically controlling for this problem weren't available, but Bromham and her colleagues were able to use more modern methods to reexamine recipes from around the world and see if they supported the earlier conclusions.

If you permit me the pun, spoiler alert: they did not. Bromham used a set of 33,750 recipes with 93 different spices from 70 cuisines on six continents. Any way you slice it, climate isn't related to a cuisine's spiciness. The risk of foodborne illness is correlated with spice use, as you would expect, but so are other diseases and conditions, including fatal car accidents, which means that people in hotter countries have lower life expectancies. But this, as Bromham says "doesn't mean that spicy food shortens your life span (or makes you crash your car)."[32] Other intervening and as-yet unexplained factors connect the two. After the publication of her work, Bromham has received emails from all over the world proposing reasons why people have spicy diets, but so far the jury is still out. Many parts of a culture arc associated with its food, including its wealth. But the moral of the story is that one should be careful about extrapolating evolutionarily derived behaviors, whether that is spiciness of food or picking a mate, from a comparison of human cultures.

The Subtler Side of Disease

The flashy effects of disease like fungus-directed ants or fearless mice get all the attention due to their novelty. But diseases exact a toll on many aspects of a human's or animal's life, often subtly, but with extremely important results. To illustrate, let's consider risk management behavior in gerbils. I realize that the phrase "risk management behavior" calls to mind studies of the way people use seatbelts or ride motorcycles, and is not usually associated with rodents. But a 2020 study[33] of Allenby's gerbil, a species that lives in the deserts in and around the Middle East, showed that the animals respond to a bacterium infection by changing their behavior in ways that could have far-reaching effects on their ecosystem.

The bacteria is a variety of *Mycoplasma*, a group of micoorganisms that can cause a range of diseases including a form of pneumonia. This variety is common among seed-eating gerbils and seems to cause deficiencies in some of their essential nutrients, though it does not outright kill the rodents. The scientists suggested[34] that the gerbils may respond to infection by changing their behavior in two different ways. First, they might compensate for the deficiency by foraging more, which exposes them to predation by owls and other night-hunting animals. Alternatively, they could simply become lethargic, which has a number of consequences, some of which are not straightforward. Lethargic gerbils are less able to escape predators because they cannot run away as quickly. That in turn means that they should spend more time looking out for predators, to give them more lead time. But spending time looking over one's shoulder, so to speak, takes away from time the animal can spend searching for seeds, further decreasing its nutritional health. Balancing the risk of being eaten against the need for food thus requires sophisticated risk management.

The researchers, who were based in Israel, the United States, and Brazil, set up large enclosures in the desert habitat of the gerbils and infected half of them with the bacteria. They then placed seeds in trays that were either close to hiding places in bushes or out in the open, and then released the tagged animals to determine their behavior. Finally, they allowed a Barn Owl into the enclosures to act as the predator, feeding the bird first to ensure it didn't simply scarf up all the gerbils in a single go. It turned out that the sick gerbils spent more time foraging, which exposed them more to the owl. They also seemed to get fewer seeds per unit of time spent searching for food, perhaps because they were, like a human suffering from the flu, simply less able to concentrate on searching. Interestingly, a chronic, long-standing infection had worse effects on the gerbils than an immediate acute infection. The upshot is that although the bacterium itself didn't kill the animals, or even keep them from apparently performing normal behaviors, it had a big effect on which gerbils were successful and, in time, which ones reproduced most successfully.

It's All in Your Head

Finally, from the You Can't Make This Stuff Up files, we have a story about sea slugs and an extreme reaction to disease. Sea slugs are the rather more glamorous cousins of the shell-less mollusks you find in your garden. Often beautifully colored, they move sinuously through the water in oceans around the world. Two species, called sacoglossan sea slugs, were recently found[35] to have an extraordinary ability: they can decapitate themselves, and then grow a completely new body, including the heart and digestive organs, from the head alone. The detached body does not respond in kind, and instead it moves around in presumed bewilderment for several days to months before it expires, a scene that surely should be incorporated into a horror film at the earliest opportunity. The heads can start eating algae, their usual food, within hours. The algae, which live inside the cells of the slug, can photosynthesize to provide the slug with energy, apparently enabling it to shortcut the usual need for nutrients. Numerous questions about this phenomenon remain; in a mastery of understatement, Sayaka Mitoh and Yoichi Ysa, the scientists who did the study, said, "The reason why the head can survive without the heart and other important organs is unclear."

Why are the slugs relevant to this chapter about behavior and disease? Many animals can regenerate body parts, such as a lizard and its tail or a salamander and a limb. They do so either as a way to avoid predators (a hungry bird ends up with a thrashing tail in its mouth while the rest of the animal escapes) or to recover from injury. The slugs, however, seem to engage in their voluntary decapitation because of—you guessed it—parasites. Sea slugs get a parasite that lives in their bodies, but it does not occupy the head. The parasite makes it difficult, if not impossible, for the slugs to reproduce, and when the head grows a new body, it is parasite-free, allowing the slug to go on to a more fulfilling future. The behavior seems a tad on the extreme side, but it goes to show that fighting disease can get you into some unexpected situations. Rudyard Kipling wrote approvingly that:

If you can keep your head when all about you
Are losing theirs and blaming it on you,
. . .
Yours is the Earth and everything that's in it

For the slugs, at least, it's not so much about keeping your head as it is about relinquishing your body, though probably that is not what Kipling had in mind.

It is fitting, I think, to conclude this book with the humble sea slug, an animal that few people have heard of and fewer still would find appealing. Yet the intricate nature of its response to infection illustrates both the complexity that exists in seemingly simple creatures and the futility of creating dichotomies among animals in terms of behavior or intelligence. Even a humble slug is capable of feats that humans cannot achieve. Those dichotomies are false whether we put humans into one category and other animals into another, or whether we allow apes, crows, and octopus into the club and exclude everything else. It is also false if we try to classify our behavior as learned and theirs as instinctive. Instead, understanding how behavior evolves allows us to celebrate the entanglement of both.

Acknowledgments

I have been thinking about the ideas in this book for many years, and have discussed versions of them with colleagues at the University of Minnesota and elsewhere. The Minnesota Center for Philosophy of Science and its Biology Interest Group provided many lively and helpful conversations through Alan Love's deft leadership. Mike Travisano, Ruth Shaw, and Mark Borrello have all influenced the way that I think about genes and behavior, and I deeply appreciate their willingness to engage with ideas and consider a variety of perspectives. My departmental colleagues in animal behavior—especially Emilie Snell-Rood, Mark Bee, and Mike Wilson—are all brilliant scientists who have listened to various arguments, talked me off various ledges, and supplied support in many ways.

I continue to be amazed at the generosity of scientists around the world and their willingness to share their thoughts, provide manuscripts, and advise me in matters outside my expertise. The ability to "cold call," or at least "cold email," someone out of the blue and receive an unfailingly enthusiastic response restores my faith in humanity, or at least that portion of humanity interested in answering obscure questions about coucals, spice use in different cuisines, or the unexceptional nature of dogs.

In 2018, I was the recipient of a Hood Fellowship from the University of Otago in Dunedin, New Zealand, where Hamish Spencer was a gracious academic host, birding and wine guide, and general discussion partner. Some of the ideas in the book were developed there. I also benefited from conversations with Lisa Matisoo-Smith of the anthropology department and members of the Department of Zoology there.

Deborah Cameron somehow always has an answer for my questions about language, feminism, and gender, and I admire her thinking and writing about all of those topics and more. Kate Nuernberger is an extraordinary poet with whom I have had many interesting conversations about writing and writers. Other scientists and scholars whose work and conversations have been influential for this book and my thinking about the evolution of behavior include Richard Wrangham, Randy Nesse, Harry Greene, Ellen Ketterson, Adrian Wenner, Jack Bradbury, and the late W. D. Hamilton.

I have always learned a great deal from my students, and Rachel Olzer deserves a special thanks for having drawn my attention, several years ago, to the difficulty of defining behavior—and why it is interesting to do so. My cricket research has benefited enormously from the effort and input of students and postdocs, especially Gita Kolluru, Nathan Bailey, Robin Tinghitella, Jessie Tanner, Susan Balenger, Justa Heinen-Kay, and the numerous undergraduates who help maintain our laboratory colonies.

Amy Cherry is a thoughtful and insightful editor who, among other things, recognized that my book's original title wasn't what I really wanted, and who has been the perfect reader for it throughout the editorial process. Huneeya Siddiqui was also a great help in polishing the manuscript and giving advice on the intricacies of publishing. My agents, Wendy Strothman and Lauren MacLeod of the Strothman Agency, have provided skillful guidance and advice. Gabriel Levin gave adept editorial assistance, and Mike van Mantgem was an able and thorough copy editor whose patience during my dithering about bird names was admirable.

Finally, my husband, John Rotenberry, has been the best field companion and source of support anyone could have asked for.

Notes

Introduction

1. Macdonald, H. "What Animals Taught Me About Being Human." *New York Times*, May 16, 2017.
2. *Washington Post*, July 7, 2019; *New York Times*, May 15, 2019; Reimann et al. (2017), "Trust Is Heritable."
3. Kresser, C. "Why Your Genes Aren't Your Destiny." Chris Kresser, February 24, 2015. https://chriskresser.com/why-your-genes-arent-your-destiny/.
4. "What Are Innate and Learned Behaviors?" ScienceNetLinks, accessed December 3, 2021. http://sciencenetlinks.com/esheets/what-are-innate-and-learned-behaviors/#:~:text=Behavior%20is%20determined%20by%20a,world%20or%20by%20being%20taught.
5. Samuels, R. "Innateness in Cognitive Science." *Trends in Cognitive Science* (2004): 136–141; Will, G. "Is the Individual Obsolete?" Minneapolis *Star Tribune*, June 16, 2019. www.startribune.com/is-the-individual-obsolete/511325222/.
6. Koshland, D. E. "Nature, Nurture and Behavior." *Science* 235 (1987): 1445.
7. Lewkowicz, D. J. "The Biological Implausibility of the Nature-Nurture Dichotomy & What It Means for the Study of Infancy." *Infancy* 16 (2011): 331–367; Spencer, J. P., Blumberg, M. S., McMurray, B., Robinson, S. R., Samuelson, L. K., and Tomblin, J. B. "Short Arms and Talking Eggs: Why We Should No Longer Abide the Nativist-Empiricist Debate." *Child Development Perspectives*, 3 (2009): 79–87; Plomin, R. "In the Nature-Nurture War, Nature Wins." *Scientific American* (2018). https://blogs.scientificamerican.com/observations/in-the-nature-nurture-war-nature-wins/.
8. Fox, J. "Let's Identify All the Zombie Ideas in Ecology!" Dynamic Ecology, February 2, 2016. Fox, J. https://dynamicecology.wordpress.com/2016/02/02/lets-identify-all-the-zombie-ideas-in-ecology/.
9. Marche, S. "The Unexamined Brutality of the Male Libido." *New York Times*, November 25, 2017.

10. Dizikes, P. "3 Questions: Evelyn Fox Keller on the Nature-Nurture Debates." Phys
    .org, December 1, 2010. https://phys.org/news/2010-12-evelyn-fox-keller-nature
    -nurture-debates.html.
11. Paul, D. B, and Brosco, J. P. *The PKU Paradox*. Johns Hopkins University Press
    (2013).
12. Goldman, D., and Landweber L. "What Is a Genome?" *PLoS Genetics*, July 2016.
    https://www.ncbi.nlm.nih.gov/pmc/articles/PMC4956268/#:~:text=Each%20
    genome%20contains%20all%20of,its%20role%20in%20the%20cell.
13. Turkheimer, E. "The Blueprint Metaphor." GHA Project, October 10, 2018. https://
    www.geneticshumanagency.org/gha/the-blueprint-metaphor/.
14. Robson, D. "Human-Like Intelligence in Animals Is Far More Common than We
    Thought." *New Scientist* (April 7, 2021); Graham, F. "Puppies Are Hardwired to
    Understand Us." *Nature* briefing, March 19, 2021; Grimm, D. "These Adorable Pup-
    pies May Help Explain Why Dogs Understand Our Body Language." *Science* news-
    letter, March 17, 2021.

1. Narwhals and the Dead Man

1. Bergner, R. M. "What Is Behavior? And Why Is It Not Reducible to Biological
   States of Affairs?" *Journal of Theoretical and Philosophical Psychology* 36 (2016):
   41–55.
2. Henriques, G., and Michalski, J. "Defining Behavior and Its Relationship to the
   Science of Psychology." *Integrative Psychological and Behavioral Science* (2020) 54:
   328–353.
3. Levitis, D. A., Lidicker, Jr, W. Z., and Freund, G. "Behavioural Biologists Do Not
   Agree on What Constitutes Behaviour." *Animal Behaviour* 78 (2009): 103–110.
4. Angier, N. "When 'What Animals Do' Doesn't Seem to Cover It." *New York Times*,
   July 20, 2009.
5. Brenner, E. D., Stahlberg, R., Mancuso, S., Vivanco, J., Baluska, F., and Van Volken-
   burgh, E. "Plant Neurobiology: An Integrated View of Plant Signaling." *Trends in
   Plant Science* 11 (2006): 413–419.
6. Giamo, C. "Watch a Flower That Seems to Remember When Pollinators Will
   Come Calling." *New York Times*, April 19, 2019; Makowski, E. "Ecuadorian Cactus
   Absorbs Ultrasound, Enticing Bats to Flowers." *The Scientist*, January 17, 2020;
   Morris, A. "A Mind Without A Brain: The Science Of Plant Intelligence Takes
   Root." *Forbes*, May 9, 2018; Pollan, M. "The Intelligent Plant." *New Yorker*,
   December 15, 2013; Taiz, L., Alkon, D., Draguhn, A., Murphy, A., Blatt, M.,
   Hawes, C., Thiel, G., and Robinson, D. G. "Plants Neither Possess nor Require
   Consciousness." *Trends in Plant Science* 24 (2019) 677–687.
7. "Linneas Sexual System." Uppsala Universitet, accessed December 3, 2021. http://
   www2.linnaeus.uu.se/online/animal/2_1.html
8. Henriques and Michalski, "Defining Behavior."
9. Ibid.
10. Reid, C. R., MacDonald, H., Mann, R. P., Marshall, J. A. R., Latty, T., and Garnier,
    S. "Decision-Making without a Brain: How an Amoeboid Organism Solves the

Two-Armed Bandit." *Journal of the Royal Society Interface* 13 (2016). http://dx.doi.org/10.1098/rsif.2016.0030.

11. Ibid.
12. "What Is Behavior?" Special Learning, Inc., accessed December 3, 2021. https://www.special-learning.com/blog/article/112; Critchfield, T. S., and Shue, E. Z. H. "The Dead Man Test: A Preliminary Experimental Analysis." *Behavior Analysis in Practice* 11 (2018): 381–384.
13. Van Dujin, M., Keijzer, F., and Franken, D. "Principles of Minimal Cognition: Casting Cognition as Sensorimotor Coordination." *Adaptive Behavior* 14 (2006): 157–170.
14. "Peter Buston Associate Professor." Buston Lab, accessed December 3, 2021. https://sites.bu.edu/bustonlab/profile/peter-buston/.
15. Peter Buston visit to author's department at the University of Minnesota, November 2019.
16. Huchard, E., English, S., Bell, M. B. V., Thavarajah, N., and Clutton-Brock, T. "Competitive Growth in a Cooperative Mammal." *Nature* 533 (2016): 532–534.
17. Ibid.
18. Slijper, E. J. "Biologic-Anatomical Investigations on the Bipedal Gait and Upright Posture in Mammals, with Special Reference to a Little Goat, Born without Forelegs." *Proceedings of the Koninklijke Nederlandse Akademie van Wetenschappen* 45 (1942): 288–295, 407–415.
19. Klopfer, P. "Behavior" (review of *Ethology of Mammals*). *Science* 165 (1969): 887.
20. Williams, T. M., Blackwell, S. B., Richter, B., Sinding, M-H. S., and Heide-Jørgensen, M. P. "Paradoxical Escape Responses by Narwhals (*Monodon monoceros*)." *Science* 358 (2017): 1328–1331.
21. Dawkins, R. *The Selfish Gene*. (1976) Oxford, Oxford University Press.
22. Hooper, R. "Chimps Have Local Culture Differences When It Comes to Eating Termites." *New Scientist*, May 28, 2020.
23. Tinbergen, N. *The Study of Instinct*. (1951) Oxford, Oxford University Press.
24. Geertz, C. "The Impact of the Concept of Culture on the Concept of Man." In *The Interpretation of Cultures*. (1966) New York, Basic Books.
25. de Waal, F. "Closer to Beast Than Angel." *Los Angeles Review of Books*, May 14, 2018.
26. Freestone, J. M. "Human Exceptionalism with a Human Face." *Areo*, April 19, 2019.
27. Bronowski, J. *The Ascent of Man*. BBC television documentary series (1973).
28. Bergner, R. M. "What Is Behavior? And Why Is It Not Reducible to Biological States of Affairs?" *Journal of Theoretical and Philosophical Psychology* 36 (2016): 41–55.
29. Balezeau, F., Wilson, B., Gallardo, G., Dick, F., Hopkins, W., Anwander, A., Friederici, A. D., Griffiths, T. D., and Petkov, C. I. "Primate Auditory Prototype in the Evolution of the Arcuate Fasciculus." *Nature Neuroscience* 23 (2020): 611–614.

2. Snakes, Spiders, Bees, and Princesses

1. Bostanchi, H., Anderson, S. C., Kami, H. G., and Papenfuss, T. J. "A New Species of Pseudocerastes with Elaborate Tail Ornamentation from Western Iran (Squamata: Viperidae)." *Proceedings of the California Academy of Sciences* 57 (2006):

443–450; Farah, T. "Meet the Snake That Hunts Birds With a Spider On Its Tail." *Discover*, April 19, 2019.

2. Durso, A. "Spider-Tailed Adders." Life is Short, but Snakes are Long, April 30, 2013. https://snakesarelong.blogspot.com/2013/04/spider-tailed-adders.html.

3. Klein, A. "Bees Force Plants to Flower Early by Cutting Holes in Their Leaves." *New Scientist*, May 21, 2020; Pashalidou, F. D., Lambert, H., Peybernes, T., Mescher, M. C., and De Moraes, C. M. "Bumble Bees Damage Plant Leaves and Accelerate Flower Production When Pollen Is Scarce." *Science* 368 (2020): 881–884.

4. Chittka, L. "The Secret Lives of Bees as Horticulturists?" *Science* 368 (2020): 824–825.

5. Cunningham, V. "Building a Nest Comes Naturally." Minneapolis *Star Tribune*, May 13, 2020.

6. Benton, M. J. "Studying Function and Behavior in the Fossil Record." *PLoS Biology* (2010) 8(3): e1000321. doi:10.1371/journal.pbio.1000321.

7. Celine, T. "*Tyrannosaurus Rex* Speed Found to Be Surprisingly Slow That Even Humans Can 'Outwalk' Them." The Science Times, April 22, 2021.

8. Emery, N. J. "Evolution of Learning and Cognition." In J. Call, G. M. Burghardt, I. M. Pepperberg, C. T. Snowdon, and T. Zentall (Eds.), *APA Handbook of Comparative Psychology: Basic Concepts, Methods, Neural Substrate, and Behavior* (2017): 237–255. American Psychological Association. https://doi.org/10.1037/0000011-012.

9. de Queiroz, A., and Wimberger, P. H. "The Usefulness of Behavior for Phylogeny Estimation: Levels of Homoplasy in Behavioral and Morphological Characters." *Evolution* 47 (1993): 46–60.

10. Bostwick, K. S. "Display Behaviors, Mechanical Sounds, and Evolutionary Relationships of the Club-Winged Manakin (*Machaeropterus deliciosus*)." *Auk* 117 (2000): 465–478; Bostwick, K. S., Riccio, M. L., and Humphries, J. M. "Massive, Solidified Bone in the Wing of a Volant Courting Bird." Biology Letters 8 (2012): 760–763.

11. Greene, H. W. "Natural History and Behavioural Homology." In *Homology*, 173–188. Novartis Foundation Symposium 222. (1999) Chichester, UK, Wiley.

12. "Trees, Not Ladders." UC Museum of Paleontology Understanding Evolution, accessed December 3, 2021. https://evolution.berkeley.edu/evolibrary/article/evo_07.

13. MacLean, P. D. "Man and His Animal Brains." *Modern Medicine* 32 (1964): 95–106.

14. Sagan, C. *The Dragons of Eden*. (1981) New York: Ballantine Books.

15. Reiner, A. "Review: An Explanation of Behavior." *Science* 250 (1990): 303–305.

16. Cesario, J., Johnson, D. J., and Eisthen, H. L. "Your Brain Is Not an Onion With a Tiny Reptile Inside." *Current Directions in Psychological Science* (2020): 2–6.

17. Ibid.

18. "Dog Breeds." American Kennel Club, accessed December 3, 2021. https://www.akc.org/dog-breeds/.

19. Email to author from Claire Wade, January 14, 2019.

20. Bateson, P. "The Active Role of Behaviour in Evolution." *Biology and Philosophy* 19 (2004): 283–298.

21. Burghardt, G. M. "Ground Rules for Dealing with Anthropomorphism." *Nature* 430 (2004): 15.

22. Darwin, C. *The Expression of the Emotions in Man and Animals*. (1872) London, John Murray.

23. Burghardt, G. M. "Darwin's Legacy to Comparative Psychology and Ethology." *American Psychologist* 64 (2009): 102–110.

24. Klopfer, P. H. "Still Largely Where Darwin Left Us." *American Psychologist* 20 (1975): 406–407.

25. Skutch, A. *A Naturalist on a Tropical Farm*. (1980) Berkeley, University of California Press.

26. de Waal, F. *Mama's Last Hug: Animal Emotions and What They Tell Us about Ourselves*. (2019) New York, W. W. Norton.

27. Giamo, C. "Praying Mantises: More Deadly Than We Knew." *New York Times*, May 14, 2020.

3. Clean-Minded Bees and Courtship Genes

1. Rothenbuhler, W. C. "Behavior Genetics of Nest Cleaning in Honey Bees. IV. Responses of F1 and Backcross Generations to Disease-Killed Brood." *American Zoologist* 4 (1964): 111–123.

2. Bastock, M. "A Gene Mutation Which Changes a Behavior Pattern." *Evolution* 10 (1956): 421–439.

3. Ibid.

4. Chandra, V., Fetter-Pruneda, I., Oxley, P. R., Ritger, A. L., McKenzie, S. K., Libbrecht, R., and Kronauer, D. J. C. "Social Regulation of Insulin Signaling and the Evolution of Eusociality in Ants." *Science* 361 (2018): 398–402; Weintraub, K. "Worker Ants: You Could Have Been Queens." *New York Times*, July 26, 2018.

5. Bell, A. M., and Robinson, G. E. "Behavior and the Dynamic Genome." *Science* 332 (2011): 1161–1162.

6. Dilger, W. C. "The Behavior of Lovebirds." *Scientific American* 206 (1962): 88–99.

7. Lewontin, R. C. "The Analysis of Variance and the Analysis of Causes." *American Journal of Human Genetics* 26 (1974): 400–411.

8. "Welcome to the Minnesota Center for Twin and Family Research!" The Minnesota Center for Twin and Family Research (MCTFR), September 21, 2017. https://mctfr.psych.umn.edu/.

9. Ibid.

10. Meffert, L. M., Hicks, S. K., and Regan, J. L. "Nonadditive Genetic Effects in Animal Behavior." *American Naturalist* 160 (2002): S198–S213.

11. Kendler, K. S., and Greenspan, R. J. "The Nature of Genetic Influences on Behavior: Lessons from 'Simpler' Organisms." *American Journal of Psychiatry* 163 (2006): 1683–1694.

12. Polderman, T., Benyamin, B., de Leeuw, C., Sullivan, P. F., van Bochoven, A., Visscher, P. M., and Posthum, D. "Meta-analysis of the Heritability of Human Traits Based on Fifty Years of Twin Studies." *Nature Genetics* 47 (2015): 702–709; Kaplan, J. "Heritability: A Handy Guide to What It Means, What It Doesn't Mean, and That Giant Meta-analysis of Twin Studies," June 1, 2015. https://scientiasalon.wordpress

.com/2015/06/01/heritability-a-handy-guide-to-what-it-means-what-it-doesnt
-mean-and-that-giant-meta-analysis-of-twin-studies/.

13. Turkheimer, E. "Three Laws of Behavior Genetics and What They Mean." *Current Directions in Psychological Science* 9 (2000): 160–164.

14. Meyer, M. N., Turley, P., and Benjamin, D. J. "Response to Charles Murray on Polygenic Scores," February 3, 2020. https://medium.com/@michellenmeyer/response -to-charles-murray-on-polygenic-scores-e768cf145cc.

15. Pan, Y., and Baker, B. S. "Genetic Identification and Separation of Innate and Experience-Dependent Courtship Behaviors in *Drosophila*." *Cell* 156 (2014): 236–248.

16. Christensen, K. D., Jayaratne, T. E., Roberts, J. S., Kardia, S. L. R., and Petty, E. M. "Understandings of Basic Genetics in the United States: Results from a National Survey of Black and White Men and Women." *Public Health Genomics* 13 (2010): 467–476.

17. Heine, S. J., Dar-Nimrod, I., Cheung, B. Y., and Proulx, T. "Essentially Biased: Why People Are Fatalistic About Genes." *Advances in Experimental Social Psychology* 55 (2017): 137–192.

18. Heine, S. J., Cheung, B. Y., and Schmalor, A. "Making Sense of Genetics: The Problem of Essentialism." Looking for the Psychosocial Impacts of Genomic Information, special report, Hastings Center Report 49 (2019): S19–S26.

19. Donovan, B. M. "Putting Humanity Back into the Teaching of Human Biology." *Studies in History and Philosophy of Biological and Biomedical Sciences* 52 (2015) 65–75.

20. Chabris, C. F., Lee, J. J., Cesarini, D., Benjamin, D. J., and Laibson, D. I. "The Fourth Law of Behavior Genetics." *Current Directions in Psychological Science* 24 (2015): 304–312.

4. Raised by Wolves—Would It Really Be So Bad?

1. Lea, S. E. G., and Osthaus, B. "In What Sense Are Dogs Special? Canine Cognition in Comparative Context." *Learning & Behavior* 46: 335–363.

2. Ducharme, J. "Your Dog Is Probably Dumber Than You Think, a New Study Says." *Time* (October 2, 2018); Morris, J. "Shocking News for Dog Lovers: Canines Aren't as Smart as We Think" *Mercury News*, October 11, 2018; CBC Radio, The Current. "New Research Suggests Dogs Aren't Exceptionally Smart," October 15, 2018. https://www .cbc.ca/radio/thecurrent/the-current-for-october-15-2018-1.4862884/new-research -suggests-dogs-aren-t-exceptionally-smart-1.4862907; Louvet, D. "This Is How Scientists Measure Intelligence In Animals." Innovet, January 12, 2019. https:// www.innovetpet.com/blogs/dogs/this-is-how-scientists-measure-intelligence-in -animals.

3. Francis, R. C. *Domesticated*. (2015) New York, W. W. Norton & Company.

4. Shipman, P. L. "The Woof at the Door." *American Scientist* 97 (2009): 286.

5. Callaway, E. "Prehistoric Genomes Reveal European Origins of Dogs." *Nature* (2013). https://doi.org/10.1038/nature.2013.14178.

6. Thalmann, O., Shapiro, B., Cui, P., Schuenemann, V. J., Sawyer, S. K., Greenfield, D. L., Germonpré, M. B., Sablin, M. V., López-Giráldez, F., Domingo-Roura,

X., Napierala, H., Uerpmann, H-P., Loponte, D. M., Acosta, A. A., Giemsch, L., Schmitz, R. W., Worthington, B., Buikstra, J. E., Druzhkova, A., Graphodatsky, A. S., Ovodov, N. D., Wahlberg, N., Freedman, A. H., Schweizer, R. M., Koepfli, K-P., Leonard, J. A., Meyer, M., Krause, J., Pääbo, S., Green, R. E., and Wayne, R. K. "Complete Mitochondrial Genomes of Ancient Canids Suggest a European Origin of Domestic Dogs." *Science* 342 (2013): 871–874.

7. van Asch, B., Zhang, A., Oskarsson, M. C. R., Klütsch, C. F. C, Amorim, A., and Savolainen, P. "Pre-Columbian Origins of Native American Dog Breeds, with Only Limited Replacement by European Dogs, Cconfirmed by mtDNA Analysis." *Proceedings of the Royal Society B* 280 (2013) http://dx.doi.org/10.1098/rspb.2013.1142.

8. Grimm, D. "Dawn of the Dog." *Science* 348 (2015): 274–279.

9. Sinding, Mikkel-Holger S., Gopalakrishnan, S., Ramos-Madrigal, J., de Manuel, M., Pitulko, V. V., Kuderna, L., Feuerborn, T. R., et al. "Arctic-Adapted Dogs Emerged at the Pleistocene–Holocene Transition." *Science* 368 (2020): 1495–1499.

10. Botigué, Laura R., Song, S., Scheu, A., Gopalan, S, Pendleton, A. L., Oetjens, M., Taravella, A. M., et al. "Ancient European Dog Genomes Reveal Continuity since the Early Neolithic." *Nature Communications* 8 (2017): 16082.

11. Francis, *Domesticated*, 25.

12. Pierotti, R., and Fogg, B. R. *The First Domestication: How Wolves and Humans Coevolved*. New Haven: Yale University Press, 2107.

13. Shipman, P. L. "The Woof at the Door." *American Scientist* 97 (2009): 286.

14. McDermott, M. T. "The Vegan Dog." *New York Times*, June 6, 2017; comments *New York Times*, March 23, 2019. www.nytimes.com/2017/06/06/well/family/the -vegan-dog.html#commentsContainer.

15. Yong, E. "How Domestication Ruined Dogs' Pack Instincts." *Atlantic*, October 16, 2017; Marshall-Pescini, S., Schwarz, J. F. L., Kostelnik, I., Virányi, Z., and Range, F. *Proceedings of the National Academy of Sciences (USA)* 114 (2017): 11793–11798.

16. Yong, "How Domestication Ruined," *Atlantic*.

17. Hansen-Wheat, C., and Temrin, H. "Intrinsic Ball Retrieving in Wolf Puppies Suggests Standing Ancestral Variation for Human-Directed Play Behavior." *iScience* (2020) doi.org/10.1016/j.isci.2019.100811; Gorman, J. "What Wolf Pups That Play Fetch Reveal About Your Dog." *New York Times*, January 20, 2020.

18. Gorman, J. "Wolf Puppies Are Adorable. Then Comes the Call of the Wild." *New York Times*, October 13, 2017; McConnell, P. B. "Canis Cousins: Unraveling Ancestral Ties." *The Bark Magazine* Spring (2004).

19. Wrangham, R. *The Goodness Paradox: The Strange Relationship Between Virtue and Violence in Human Evolution*. (2019) New York, Pantheon.

20. Darwin, C. *The Variation of Plants and Animals Under Domestication*. (1868) London, John Murray.

21. Dugatkin, L. A., and Trut, L. *How to Tame a Fox (and Build a Dog)*. (2017) Chicago, University of Chicago Press.

22. Ibid.

23. Wilkins, A. S., Wrangham, R. W., and Fitch, W. T. "The 'Domestication Syndrome' in Mammals: A Unified Explanation Based on Neural Crest Cell Behavior and Genetics." *Genetics* 197 (2014): 795–808.

24. Lord, K. A., Larson, G., Coppinger, R. P., Karlsson, E. K. "The History of Farm Foxes Undermines the Animal Domestication Syndrome." *Trends in Ecology & Evolution* 35 (2020): 125–136.

25. Ibid.

26. Johnson, A. "Fact or Fiction: Foxes Were Brought to Catalina Island by the Native People." Online newsletter, Catalina Island Conservancy (2013) Issue 30.

27. Lord et al. "History of Farm Foxes."

28. Hansen Wheat, C., Fitzpatrick, J., Rogell, B., and Temrin, H. "Behavioural Correlations of the Domestication Syndrome Are Decoupled in Modern Dog Breeds." *Nature Communications* 10, no. 1 (2019).

29. Darwin, C. *The Origin of Species.* (1859) London, John Murray.

30. Gábor, A., Gácsi, M., Szabó, D., Miklósi, A., Kubinyi, E., and Andics, A. "Multilevel fMRI Adaptation for Spoken Word Processing in the Awake Dog Brain." *Scientific Reports* 10 (2020): 11968.

31. Lea and Osthaus, "In What Sense."

32. Email to author, August 12, 2020.

33. Yong, E. "Can Dogs Smell Their 'Reflections'?" *Atlantic*, August 17, 2017.

34. Ibid.

35. "Join the World's Largest Pet Citizen Science Project." Darwin's Ark, accessed December 3, 2021. https://darwinsark.org/.

36. Ibid.

37. "Find the Genius in Your Dog." Dognition, accessed December 3, 2021. https://www.dognition.com/.

38. Woods, V., and Hare, B. "Why Are There So Many Books About Dogs?" *New York Times*, April 29, 2019.

39. Nagasawa, M., Mitsui, S., En, S., Ohtani, N., Ohta, M., Sakuma, Y., Onaka, T., Mogi, K., and Kikusui, K. "Oxytocin-Gaze Positive Loop and the Coevolution of Human-Dog Bonds." *Science* 348 (2015): 333–336; MacLean, E. L., and Hare, B. "Dogs Hijack the Human Bonding Pathway." *Science* 348 (2015): 280–281.

5. Wild-Mannered

1. "Only in Brooklyn." Why Evolution Is True, August 2, 2017. https://whyevolutionistrue.com/2017/08/02/only-in-brooklyn/.

2. Lawler, A. *Why Did the Chicken Cross the World? The Epic Saga of the Bird that Powers Civilization.* (2014) New York, Atria Books.

3. Driscoll, C. A., Macdonald, D. W., and O'Brien, S. J. "From Wild Animals to Domestic Pets, an Evolutionary View of Domestication." *Proceedings of the National Academy of Sciences (USA)* 106 (2009): 9971–9978.

4. Ottoni, C., Van Neer, W., De Cupere, B., Daligault, J., Guimaraes, S., Peters, J., Spassov, N., Prendergast, M. E., Boivin, N., Morales-Muñiz, A., Bălăşescu, A., Becker, C., Benecke, N., Boroneant, A., Buitenhuis, H., Chahoud, J., Crowther, A., Llorente, L., Manaseryan, N., Monchot, H., Onar, V., Osypińska, M., Putelat, O., Quintana Morales, E. M., Studer, J., Wiere, U., Decorte, R., Grange, T., and Geigl, E-M. "The Palaeogenetics of Cat Dispersal in the Ancient World." *Nature Ecology &*

Evolution 1 (2017): 0139; Newitz, A. "Cats Are an Extreme Outlier among Domestic Animals." *Ars Technica*, June 19, 2017. https://arstechnica.com/science/2017/06/cats-are-an-extreme-outlier-among-domestic-animals

5. Driscoll, "From Wild Animals."

6. Hu, Y., Hu, S., Wang, W., Wu, X., Marshall, F. B., Chen, X., Hou, H., and Wang, C. "Earliest Evidence for Commensal Processes of Cat Domestication." *Proceedings of the National Academy of Sciences* 111 (2014): 116–120.

7. Hu et al., "Earliest Evidence."

8. Montague, M. J., Gang Li, B. G., Razib Khan, B. L., Aken, S. M. J., Searle, P. M., et al. "Comparative Analysis of the Domestic Cat Genome Reveals Genetic Signatures Underlying Feline Biology and Domestication." *Proceedings of the National Academy of Sciences* 111 (2014): 17230–17235.

9. Berteselli, G. V., Regaiolli, B., Normando, S., De Mori, B., Zaborra, C. A., and Spiezio, C. "European Wildcat and Domestic Cat: Do They Really Differ?" *Journal of Veterinary Behavior* 22 (2017): 35–40.

10. Achterberg, P. "The Reason Why Cats Don't behave, While Dogs Do." *Blue Skies & Happy Clouds*, July 20, 2020. https://medium.com/blue-skies-happy-clouds/the-reason-why-cats-dont-behave-while-dogs-do-b59446198c70.

11. Zak, P. "Dogs (and Cats) Can Love." *Atlantic*, April 22, 2014.

12. Wang, M-S., Thakur, M., Peng, M-S., Jiang, Y., Frantz, L. A. F., Li, M., Zhang, J-J., et al. "863 Genomes Reveal the Origin and Domestication of Chicken." *Cell Research* (2020): 693–701.

13. "Police Officer Raiding Illegal Cockfight Gets Killed by Rooster." BBC News, October 28, 2020. https://www.bbc.com/news/world-asia-54715327.

14. Lawler, A. *Why Did the Chicken Cross the World?*

15. Gering, E., Incorvaia, D., Henriksen, R., Conner, J., Getty, T., and Wright, D. "Getting Back to Nature: Feralization in Animals and Plants." *Trends in Ecology & Evolution* 34 (2019): 1137–1151; Johnsson, M., Gering, E., Willis, P., Lopez, S., Van Dorp, L., Hellenthal, G., Henriksen, R., Friberg, U., and Wright, D. "Feralisation Targets Different Genomic Loci to Domestication in the Chicken." *Nature Communications* 7 (2016): 12950.

16. Lord, E., Collins, C., deFrance, S., LeFebvre, M. J., Pigière, F., Eeckhout, P., Erauw, C., et al. "Ancient DNA of Guinea Pigs (*Cavia* spp.) Indicates a Probable New Center of Domestication and Pathways of Global Distribution." *Scientific Reports* 10 (June 2020): 8901.

17. Engber, Daniel. "Test-Tube Piggies." Slate, June 18, 2012. https://slate.com/technology/2012/06/human-guinea-pigs-and-the-history-of-the-iconic-lab-animal.html.

18. Ibid.

19. Ibid.

20. Zimmermann, T. D., Kaiser, S., Hennessy, M. B., and Sachser, N. "Adaptive Shaping of the Behavioural and Neuroendocrine Phenotype during Adolescence." *Proceedings of the Royal Society B: Biological Sciences* 284 (2017): 20162784; Kaiser, S., Hennessy, M. B., and Sachser, N. "Domestication Affects the Structure, Development and Stability of Biobehavioural Profiles." *Frontiers in Zoology* 12 (2015): S19.

21. Brust, V., and Guenther, A. "Domestication Effects on Behavioural Traits and Learning Performance: Comparing Wild Cavies to Guinea Pigs." *Animal Cognition* 18 (2015): 99–109.

22. Offord, S. "Are Honey Bees Domesticated Livestock?" Keeping Backyard Bees, July 21, 2017. https://www.keepingbackyardbees.com/honey-bees-domesticated-livestock/; "Bees Are Not Domestic Animals." Buzz Beekeeping Supplies, accessed December 3, 2021. http://buzzbeekeepingsupplies.com/bees-are-not-domestic-animals/.

23. Seeley, T. D. "Who Were the Geniuses Who First Domesticated the Wild Honey Bee?" *Literary Hub*, May 24, 2019. https://lithub.com/who-were-the-geniuses-who-first-domesticated-the-wild-honey-bee/.

24. Driscoll et al. "From Wild Animals."

25. Seeley, T. "Who Were the Geniuses."

26. Ibid.

27. Buzz Beekeeping Supplies. "Bees Are Not Domestic Animals."

28. Oldroyd, B. P. "Corrigendum." *Molecular Ecology* 22 (2013): 1483; Oldroyd, B. P. "Domestication of Honey Bees Was Associated with Expansion of Genetic Diversity." *Molecular Ecology* 21 (2012): 4409–4411.

29. Buzz Beekeeping Supplies. "Bees Are Not Domestic Animals."

30. Mehlhorn, J., and Petow, S. "Smaller Brains in Laying Hens: New Insights into the Influence of Pure Breeding and Housing Conditions on Brain Size and Brain Composition." *Poultry Science* 99 (2020): 3319–3327.

31. Ibid.

32. Henriksen, R., Johnsson, M., Andersson, L., Jensen, P., and Wright, D. "The Domesticated Brain: Genetics of Brain Mass and Brain Structure in an Avian Species." *Scientific Reports* 6 (2016): 34031.

33. Agnvall, B., Ali, A., Olby, S., and Jensen, P. "Red Junglefowl (*Gallus gallus*) Selected for Low Fear of Humans Are Larger, More Dominant and Produce Larger Offspring." *Animal* 8 (2014): 1498–1505.

34. Dudde, A., Krause, E. T., Matthews, L. R., and Schrader, L. "More Than Eggs—Relationship Between Productivity and Learning in Laying Hens." *Frontiers in Psychology* (2018): 2000.

35. Henriksen et al. "The Domesticated Brain."

6. The Anxious Invertebrate

1. Bacqué-Cazenave, J., Bharatiya, R., Barrière, G., Delbecque, J-P., Bouguiyoud, N., Di Giovanni, D., Cattaert, D., and De Deurwaerdère, P. "Serotonin in Animal Cognition and Behavior." *International Journal of Molecular Sciences* 21 (2020): 1649; Bacqué-Cazenave, J., Cattaert, D., Delbecque, J-P., and Fossat, P. "Social Harassment Induces Anxiety-like Behaviour in Crayfish." *Scientific Reports* 7 (2017): 39935.

2. Zimmer, C. *She Has Her Mother's Laugh*. (2018) New York, Dutton, 45.

3. Gordon, J. "What Can Animals Tell Us about Mental Illnesses?" National Institute of Mental Health, October 21, 2019. https://www.nimh.nih.gov/about/director/messages/2019/what-can-animals-tell-us-about-mental-illnesses.

4. Braitman, L. *Animal Madness*. (2014) New York, Simon & Schuster, 3.

5. Korneliussen, I. "Can Wild Animals Have Mental Illnesses?" Sciencenorway.no, June 24, 2015. https://sciencenorway.no/animal-behaviour-animal-kingdom -domestication/can-wild-animals-have-mental-illnesses/1417928.

6. "15 Things to Know about Mental Disorders in Animals." Online Psychology Degree Guide, accessed December 3, 2021. https://www.onlinepsychologydegree .info/lists/information-mental-disorders-in-animals/.

7. Dasgupta, S. "Many Animals Can Become Mentally Ill." BBC, September 9, 2015. http://www.bbc.com/earth/story/20150909-many-animals-can-become -mentally-ill.

8. Nesse, R. *Good Reasons for Bad Feelings*. (2019) New York, Dutton: 15.

9. Price, J. "The Dominance Hierarchy and the Evolution of Mental Illness." *The Lancet* 290 (1967): 243–246.

10. Ibid.

11. "Mental Disorders." World Health Organization, November 28, 2019. https:// www.who.int/news-room/fact-sheets/detail/mental-disorders.

12. Nesse, R. *Good Reasons for Bad Feelings*.

13. Ibid

14. Durisko, Z., Mulsant, B. H., McKenzie, K., and Andrews, P. W. "Using Evolutionary Theory to Guide Mental Health Research." *Canadian Journal of Psychiatry* 61 (2016): 159–165.

15. "Obsessive-Compulsive Disorder (OCD)." National Institute of Mental Health, accessed December 3, 2021. https://www.nimh.nih.gov/health/statistics/ obsessive-compulsive-disorder-ocd.

16. Karlsson, E. "What Clues Does Your Dog's Spit Hold for Human Mental Health?" Karlsson Lab, accessed December 3, 2021. https://karlssonlab.org/about-darwins -dogs/.

17. Dodman, N. H., Karlsson, E. K., Moon-Fanelli, A., Galdzicka, M., Perloski, M., Shuster, L., Lindblad-Toh, K., and Ginns, E. I. "A Canine Chromosome 7 Locus Confers Compulsive Disorder Susceptibility." *Molecular Psychiatry* 15, no. 1, January 2010: 8–10.

18. From presentation sent to author by Elinor Karlsson, January 29, 2019.

19. Wolmarans, D. W., Scheepers, I. M., Stein, D. J., and Harvey, B. H. "*Peromyscus maniculatus bairdii* as a Naturalistic Mammalian Model of Obsessive-Compulsive Disorder: Current Status and Future Challenges." *Metabolic Brain Disease* 33 (2018): 443–455.

20. Dodman, N. *Pets on the Couch*. (2016) New York, Atria Books.

21. Fossat, P., Bacque-Cazenave, J., De Deurwaerdere, P., Delbecque, J-P., and Cattaert, D. "Anxiety-like Behavior in Crayfish Is Controlled by Serotonin." *Science* 344 (2014): 1293–1297; Bacqué-Cazenave et al., "Serotonin in Animal Cognition and Behavior"; Bacqué-Cazenave et al., "Social Harassment Induces."

22. Bacqué-Cazenave, J., Berthomieu, M., Cattaert, D., Fossat, P., and Delbecque, J-P. "Do Arthropods Feel Anxious during Molts?" *Journal of Experimental Biology* 222 (2018): jeb.186999.

23. Ibid.

24. Jacob, F. "Evolution and Tinkering." *Science* 196 (1977): 1161–1166.
25. Braitman, L. *Animal Madness*, 208.
26. Reardon, S. "How Evolution Has Shaped Mental Illness." *Nature* 551 (2017): 15–16.
27. Scott, D., and Tamminga, C. A. "Effects of Genetic and Environmental Risk for Schizophrenia on Hippocampal Activity and Psychosis-like Behavior in Mice." *Behavioural Brain Research* 339 (2018): 114–123.
28. Song, J. H. T., Lowe, C. B., and Kingsley, D. M. "Characterization of a Human-Specific Tandem Repeat Associated with Bipolar Disorder and Schizophrenia." *American Journal of Human Genetics* 103 (2018): 421–430.
29. Bielecka, K., and Marcinów, M. "Mental Misrepresentation in Non-Human Psychopathology." *Biosemiotics* 10 (2017): 195–210.
30. Ibid.
31. Dodman, N. *Pets on the Couch*; Dodman, N. H. *The Dog Who Loved Too Much: Tales, Treatments and the Psychology of Dogs.* (1996) New York, Bantam.
32. Dodman, N. *Pets on the Couch*: 261.
33. Ibid.
34. Janiak, M. C., Pinto, S. L., Duytschaever, G., Carrigan, M. A., and Melin, A. D. "Genetic Evidence of Widespread Variation in Ethanol Metabolism among Mammals: Revisiting the 'Myth' of Natural Intoxication." *Biology Letters* 16 (2020): 20200070.
35. Ibid.
36. Ibid.

7. Dancing Cockatoos and Thieving Gulls

1. Cassella, C. "Gulls Work Out the Timing of School Lunch Breaks So They Can Steal Food." ScienceAlert, November 13, 2020. https://www.sciencealert.com/gulls-in-the-uk-have-figured-out-when-schools-are-on-lunch-break-so-they-can-steal-food.
2. Carey, B. "Alex, a Parrot Who Had a Way With Words, Dies." *New York Times*, September 10, 2007.
3. Sample, I. "Cockatoo Choreographs His Own Dance Moves, Researchers Believe." *Guardian*, July 8, 2019.
4. Patel, A. D., Iversen, J. R., Bregman, M. R., and Schulz, I. "Experimental Evidence for Synchronization to a Musical Beat in a Nonhuman Animal." *Current Biology* 19 (2009): 827–830; Jao Keehn, J., Iversen, J. R., Schulz, I., and Patel, A. D. "Spontaneity and Diversity of Movement to Music Are Not Uniquely Human." *Current Biology* 29 (2019): R621–622.
5. Ibid
6. Kim, S. E. "Why Australia's Trash Bin–Raiding Cockatoos Are the 'Punks of the Bird World.'" *Smithsonian Magazine*, July 22, 2021; Klump, B. C., Martin, J. M., Wild, S., Hörsch, J. K., Major, R. E., and Aplin, L. M. "Innovation and Geographic Spread of a Complex Foraging Culture in an Urban Parrot." *Science* 373 (2021): 456–460.
7. Hunt, G. R. "Manufacture and Use of Hook-Tools by New Caledonian Crows." *Nature* 379 (1996): 249–251.

8. Gruber, R., Schiestl, M., Boeckle, M., Frohnwieser, A., Miller, R., Gray, R. D., Clayton, N. S., and Taylor, A. H. "New Caledonian Crows Use Mental Representations to Solve Metatool Problems." *Current Biology* 29 (2019): 686–692; McCoy, D. E., Schiestl, M., Neilands, P., Hassall, R., Gray, R. D., and Taylor, A. H. "New Caledonian Crows Behave Optimistically after Using Tools." *Current Biology* 29 (2019): 2737–2742.

9. McCoy et al. "New Caledonian Crows Behave Optimistically after Using Tools."

10. Boeckle, M., and Clayton, N. S. "A Raven's Memories Are for the Future." *Science* 357, no. 6347 (July 14, 2017): 126–127; Kabadayi, C., and Osvath, M. "Ravens Parallel Great Apes in Flexible Planning for Tool-Use and Bartering." *Science* 357, no. 6347, July 14, 2017: 202–204.

11. Laumer, I. B., Jelbert, S. A., Taylor, A. H., Rössler, T., and Auersperg, A. M. I. "Object Manufacture Based on a Memorized Template: Goffin's Cockatoos Attend to Different Model Features." *Animal Cognition* 24 (2021): 457–470; Auersperg, A. M. I., Köck, C., Pledermann, A., O'Hara, M., and Huber, L. "Safekeeping of Tools in Goffin's Cockatoos, *Cacatua goffiniana*." *Animal Behaviour* 128 (2017): 125–133; Rössler, T., Mioduszewska, B., O'Hara, M., Huber, L., Prawiradilaga, D. M., and Auersperg, A. M. I. "Using an Innovation Arena to Compare Wild Caught and Laboratory Goffin's Cockatoos." *Scientific Reports* 10 (2020).

12. Rössler et al. "Using an Innovation Arena."

13. Ibid.

14. Bandini, E., Motes-Rodrigo, A., Steele, M. P., Rutz, C., and Tennie, C. "Examining the Mechanisms Underlying the Acquisition of Animal Tool Behaviour." *Biology Letters* 16 (2020): 20200122.

15. Fine, M. L., Horn, M. H., and Cox, B. "*Acanthonus armatus*, a Deep-Sea Teleost Fish with a Minute Brain and Large Ears." *Proceedings of the Royal Society B: Biological Sciences* 230 (1987): 257–265.

16. "Encephalization Quotient." Psychology Wiki. Accessed November 29, 2020. https://psychology.wikia.org/wiki/Encephalization_quotient.

17. Cairó, O. "External Measures of Cognition." *Frontiers in Human Neuroscience* 5 (2011): 108.

18. Bastos, A. P. M., and Taylor, A. H. "Macphail's Null Hypothesis of Vertebrate Intelligence: Insights from Avian Cognition." *Frontiers in Psychology* 11 (2020): 1692.

19. Jarvis, E. D., Güntürkün, O., Bruce, L., Csillag, A., Karten, H., Kuenzel, W., Medina, L. et al. "Avian Brains and a New Understanding of Vertebrate Brain Evolution." *Nature Reviews Neuroscience* 6 (2005): 151–159.

20. Ibid.

21. Gorman, J. "Crows Clever Enough to Learn a Shell Game." *New York Times*, February 29, 2016.

22. Lefebvre, L., Whittle, P., Lascaris, E., and Finkelstein, A. "Feeding Innovations and Forebrain Size in Birds." *Animal Behaviour* 53 (1997): 549–560.

23. Nicolakakis, N., and Lefebvre, L. "Forebrain Size and Innovation Rate in European Birds: Feeding, Nesting and Confounding Variables." *Behaviour* 137 (2000): 1415–1429.

24. Ducatez, S., Sol, D., Sayol, F., and Lefebvre, L. "Behavioural Plasticity Is Associated

with Reduced Extinction Risk in Birds." *Nature Ecology & Evolution* 4, no. 6 (June 2020): 788–793.

25. Ibid.; Elbein, A. "These Birds Eat Fire, or Close to It, to Live Another Day." *New York Times*, April 28, 2020.

26. Sayol, F., Lapiedra, O., Ducatez, S., and Sol, D. "Larger Brains Spur Species Diversification in Birds." *Evolution* 73 (2019): 2085–2093.

27. Sol, D. "Revisiting the Cognitive Buffer Hypothesis for the Evolution of Large Brains." *Biology Letters* 5 (2009): 130–133.

28. Sayol, F., Maspons, J., Lapiedra, O., Iwaniuk, A. N., Székely, T., and Sol, D. "Environmental Variation and the Evolution of Large Brains in Birds." *Nature Communications* 7 (2016): 13971.

29. Jiménez-Ortega, D., Kolm, N., Immler, S., Maklakov, A. A., and Gonzalez-Voyer, A. "Long Life Evolves in Large-Brained Bird Lineages." *Evolution* 74 (2020): 2617–2628.

30. Yong, E. "Can Dogs Smell Their 'Reflections'?" *The Atlantic*, August 17, 2017.

31. Kohda, M., HottaI, T., Takeyama, T., Awata, S., Tanaka, H., Asai, J-Y., and Jordan, A. L. "If a Fish Can Pass the Mark Test, What Are the Implications for Consciousness and Self-Awareness Testing in Animals?" *PLoS Biology* 17(2): e3000021.

32. Wilke, C. "The Mirror Test Peers into the Workings of Animal Minds." *The Scientist*, February 21, 2019.

33. Schulze-Makuch, D. "The Naked Mole-Rat: An Unusual Organism with an Unexpected Latent Potential for Increased Intelligence?" *Life* 9 (2019): 76.

34. Ibid.

35. Quora. "Why Is There Only One Intelligent Species? Why Aren't Other Species Intelligent?" Accessed September 14, 2017. https://www.quora.com/Why-is-there-only-one-intelligent-species-Why-aren%E2%80%99t-other-species-intelligent?share=1.

36. Snell-Rood, E. C., and Wick, N. "Anthropogenic Environments Exert Variable Selection on Cranial Capacity in Mammals." *Proceedings of the Royal Society B* (2013) 280: 20131384.

8. A Soft Spot for Hard Creatures

1. Mattila, H. R., Otis, G. W., Nguyen, L. T. P., Pham, H. D., Knight, O. M., and Phan, N. T. "Honey Bees (*Apis cerana*) Use Animal Feces as a Tool to Defend Colonies against Group Attack by Giant Hornets (*Vespa soror*)." *PLoS ONE* 15 (2020): e0242668.

2. Ibid.

3. Carrington, D. "Honey Bees Use Animal Poo to Repel Giant Hornet Attacks." *Guardian*, December 9, 2020.

4. Coyne, J. "Tool-Using Ants Build Siphons to Wick Sugar Water Out of Containers, Keeping Them from Drowning." *Why Evolution Is True* (blog), October 11, 2020. https://whyevolutionistrue.com/2020/10/11/tool-using-ants-build-siphons-to-wick-sugar-water-out-of-containers-so-they-dont-drown/; Zhou, A., Du, Y., and Chen, J. "Ants Adjust Their Tool Use Strategy in Response to Foraging Risk." *Functional Ecology* 34, no. 12 (December 2020): 2524–2535.

5. Ibid.
6. Godfrey-Smith, P. *Metazoa: Animal Life and the Birth of the Mind*. (2020) New York, Farrar, Straus and Giroux, 81.
7. Shpigler, H. Y., Saul, M. C., Corona, F., Block, L., Ahmed, A. C., Zhao, S. D., and Robinson, G. E. "Deep Evolutionary Conservation of Autism-Related Genes." *Proceedings of the National Academy of Sciences* 114 (2017): 9653–9658.
8. Ibid.
9. Pennisi, E. "Antisocial Bees Share Genetic Profile with People with Autism." *Science* (July 31, 2017). https://doi.org/10.1126/science.aan7184.
10. Dvorsky, G. "Bees Can Learn Symbols Associated with Counting, New Experiment Suggests." Gizmodo, June 5, 2019. https://gizmodo.com/bees-can-learn-symbols-associated-with-counting-new-ex-1835245766; Howard, S. R., Avarguès-Weber, A., Garcia, J. E., Greentree, A. D., and Dyer, A. G. "Symbolic Representation of Numerosity by Honeybees (*Apis mellifera*): Matching Characters to Small Quantities." *Proceedings of the Royal Society B: Biological Sciences* 286 (2019): 20190238.
11. Ibid.
12. Davies, R., Gagen, M. H., Bull, J. C., and Pope, E. C. "Maze Learning and Memory in a Decapod Crustacean." *Biology Letters* 15 (2019): 20190407.
13. Pan, Y., and Baker, B. S. "Genetic Identification and Separation of Innate and Experience-Dependent Courtship Behaviors in *Drosophila*." *Cell* 156 (2014): 236–248.
14. Nash, O. *Good Intentions*. (1942) New York, Little, Brown.
15. Godfrey-Smith, P. *Other Minds: The Octopus, the Sea, and the Deep Origins of Consciousness*. (2016) New York, Farrar, Straus and Giroux.
16. Ibid., 52.
17. Kuba, M. J., Gutnick, T., and Burghardt, G. M. "Learning from Play in Octopus." In *Cephalopod Cognition*, edited by Anne-Sophie Darmaillacq, Ludovic Dickel, and Jennifer Mather, 57–71. (2014) Cambridge: Cambridge University Press.
18. Ibid, 66.
19. Godfrey-Smith, P. *Metazoa*, 142.
20. Amodio, P., Boeckle, M., Schnell, A. K., Ostojíc, L., Fiorito, G., and Clayton, N. S. "Grow Smart and Die Young: Why Did Cephalopods Evolve Intelligence?" *Trends in Ecology & Evolution* 34 (2019): 45–56.
21. Ibid.
22. Sampaio, E., Seco, M. C., Rosa, R., and Gingins, G. "Octopuses Punch Fishes during Collaborative Interspecific Hunting Events." *Ecology* 102 (2021): e03266; Dockrill, P. "Octopuses Observed Punching Fish, Perhaps Out of Spite, Scientists Say." ScienceAlert, December 21, 2020. https://www.sciencealert.com/octopuses-observed-punching-fish-perhaps-out-of-spite-scientists-say; Frishberg, H. "Octopuses Spite-Punch Fish, Who 'Don't Like It,' Study Finds." *New York Post*, December 23, 2020. https://nypost.com/2020/12/23/octopuses-spite-punch-fish-who-dont-like-it-study/; Preston, E. "Eight-Armed Underwater Bullies: Watch Octopuses Punch Fish." *New York Times* (December 24, 2020). https://www.nytimes.com/2020/12/24/science/octopus-punch-fish.html.
23. Zuk, M. *Sex on Six Legs*. (2013) New York, Houghton Mifflin Harcourt.

24. Darwin, C. *The Descent of Man and Selection in Relation to Sex.* (1871) London, John Murray.

25. Eberhard, W. G. "Are Smaller Animals Behaviourally Limited? Lack of Clear Constraints in Miniature Spiders." *Animal Behaviour* 81 (2011): 813–823.

26. Eberhard, W. G., and Wcislo, W. T. "Plenty of Room at the Bottom?" *American Scientist* 100 (June 2012): 226–233.

27. Sayol, F., Collado, M. Á., Garcia-Porta, J., Seid, M. A., Gibbs, J., Agorreta, A., San Mauro, D., Raemakers, I., Sol, D., and Bartomeus, I. "Feeding Specialization and Longer Generation Time Are Associated with Relatively Larger Brains in Bees." *Proceedings of the Royal Society B: Biological Sciences* 287 (2020): 20200762; Preston, E. "Meet a Bee with a Very Big Brain." *New York Times*, July 15, 2020.

28. Chittka, L., and Niven, J. "Are Bigger Brains Better?" *Current Biology* 19 (2009): R995–1008; Niven, J. E., and Chittka, L. "Evolving Understanding of Nervous System Evolution." *Current Biology* 26 (2016): R937–941.

29. Ibid.

30. Zuk, M., Rotenberry, J. T., and Tinghitella, R. M. "Silent Night: Adaptive Disappearance of a Sexual Signal in a Parasitized Population of Field Crickets." *Biology Letters* 2 (2006): 521–524.

31. Ibid.

32. Bailey, N. W., and Zuk, M. "Field Crickets Change Mating Preferences Using Remembered Social Information." *Biology Letters* 5 (2009): 449–451.

33. Snell-Rood, E. C., and Papaj, D. R. "Patterns of Phenotypic Plasticity in Common and Rare Environments: A Study of Host Use and Color Learning in the Cabbage White Butterfly *Pieris rapae*." *American Naturalist* 173 (2009): 615–631.

34. Cross, F. R., Carvell, G. A., Jackson, R. R., and Grace, R. C. "Arthropod Intelligence? The Case for *Portia*." *Frontiers in Psychology* 11 (2020): 568049.

35. Shettleworth, S. J. "Clever Animals and Killjoy Explanations in Comparative Psychology." *Trends in Cognitive Sciences* 14 (2010): 477–481.

36. Thorndike, E. L. "Animal Intelligence: An Experimental Study of the Associative Processes in Animals." *Psychological Review Monograph Supplement* 2 (1898): 1–109.

37. Dennett, D. C. "Animal Consciousness: What Matters and Why." *Social Research* 62 (1995): 691–710.

38. Godfrey-Smith, P. *Other Minds*.

39. Hernandez-Lallement, J., van Wingerden, M., and Kalenscher, T. "Towards an Animal Model of Callousness." *Neuroscience and Biobehavioral Reviews* 91 (2018): 121–129.

9. Talking with the Birds and the Bees. And the Monkeys.

1. Bergner, R. M. "What Is Behavior? And So What?" *New Ideas in Psychology* 29 (2011): 147–155.

2. Wilson, A. N. "Why I Believe Again." *New Statesman*, April 2, 2009.

3. Wade, N. "Early Voices: The Leap to Language." *New York Times*, July 15, 2003.

4. Bickerton, D. "Where Language Comes From." Harvard University Press Blog (blog), January 29, 2014. https://harvardpress.typepad.com/hup_publicity/2014/01/more-than-nature-needs-derek-bickerton.html.

5. Wrangham, R. *The Goodness Paradox*.

6. Ibid., 166.

7. Hockett, C. F., and Ascher, R. "The Human Revolution [and Comments and Reply]." *Current Anthropology* 5 (1964): 135–168.

8. Ibid.

9. Wilson, A. N. "Why I Believe Again."

10. Balezeau, F., Wilson, B., Gallardo, G., Dick, F., Hopkins, W., Anwander, A., Friederici, A. D., Griffiths, T. D., and Petkov, C. I. "Primate Auditory Prototype in the Evolution of the Arcuate Fasciculus." *Nature Neuroscience* 23 (2020): 611–614.

11. Boë, L-J., Sawallis, T. R., Fagot, J., Badin, P., Barbier, G., Captier, G., Ménard, L., Heim, J-L., and Schwartz, J-L. "Which Way to the Dawn of Speech?: Reanalyzing Half a Century of Debates and Data in Light of Speech Science." *Science Advances* 5 (2019): eaaw3916.

12. Pereira, A. S., Kavanagh, E., Hobaiter, C., Slocombe, K. E., and Lameira, A. R. "Chimpanzee Lip-Smacks Confirm Primate Continuity for Speech-Rhythm Evolution." *Biology Letters* 16 (2020): 20200232.

13. Corballis, M. C. "Crossing the Rubicon: Behaviorism, Language, and Evolutionary Continuity." *Frontiers in Psychology* 11 (2020): 653; Corballis, M. C. "Language Evolution: A Changing Perspective." *Trends in Cognitive Sciences* 21 (April 2017): 229–236; Corballis, M. C. "Mental Time Travel, Language, and Evolution." *Neuropsychologia* 134 (2019): 107202.

14. Graham, K. E., Hobaiter, C., Ounsley, J., Furuichi, T., and Byrne, R. W. "Bonobo and Chimpanzee Gestures Overlap Extensively in Meaning." *PLoS Biology* 16 (2018): e2004825.

15. "The Great Ape Dictionary." The Great Ape Dictionary, accessed December 3, 2021. https://greatapedictionary.ac.uk/.

16. Novack, M. A., and Waxman, S. "Becoming Human: Human Infants Link Language and Cognition, but What about the Other Great Apes?" *Philosophical Transactions of the Royal Society B: Biological Sciences* 375 (2020): 20180408.

17. "FAQ: How Did Language Begin?" Linguistic Society of America, accessed December 3, 2021. https://www.linguisticsociety.org/resource/faq-how-did-language-begin.

18. Shettleworth, S. J. "Clever Animals and Killjoy Explanations."

19. Godfrey-Smith, P. *Other Minds*, 143.

20. Corballis, M. C. "Language Evolution: A Changing Perspective."

21. Vernes, S. C., and Wilkinson, G. S. "Behaviour, Biology and Evolution of Vocal Learning in Bats." *Philosophical Transactions of the Royal Society B: Biological Sciences* 375 (2019): 20190061.

22. Ibid.

23. Piattelli, M. "'Language Is Our Rubicon': Friedrich Max Müller's Quarrel with Hensleigh Wedgwood." *Publications of the English Goethe Society* 85 (2016): 98–109.

24. Suzuki, T. N., and Zuberbühler, K. "Animal Syntax." *Current Biology* 29 (2019): R669–671.

25. Kershenbaum, A., Demartsev, V., Gammon, D. E., Geffen, E., Gustison, M. L., Ilany, A., and Lameira, A. R. "Shannon Entropy as a Robust Estimator of Zipf's Law in Animal Vocal Communication Repertoires." *Methods in Ecology and Evolution* 12 (2020): 553–564.

26. Speck, B., Seidita, S., Belo, S., Johnson, S., Conley, C., Desjonquères, C., and Rodríguez, R. L. "Combinatorial Signal Processing in an Insect." *American Naturalist* 196
(2020): 406–413.
27. Ibid.
28. Goldman, J. G. "Ronan the Sea Lion Dances to the Backstreet Boys. So What?"
Scientific American, April 4, 2013.
29. Araya-Salas, M. "Is Birdsong Music? Evaluating Harmonic Intervals in Songs of a
Neotropical Songbird." *Animal Behaviour* 84 (2012): 309–313.
30. Gábor, A., Gácsi, M., Szabó, D., Miklósi, A., Kubinyi, E., and Andics, A. "Multilevel
fMRI Adaptation for Spoken Word Processing in the Awake Dog Brain." *Scientific
Reports* 10 (2020): 11968.
31. Magyari, L., Huszár, Zs., Turzó, A., and Andics, A. "Event-Related Potentials
Reveal Limited Readiness to Access Phonetic Details during Word Processing in
Dogs." *Royal Society Open Science* 7 (2020): 200851.
32. Woodyatt, A. "Sorry, Folks—Your Dog Doesn't Really Know What You're Talking
About." CNN Philippines, December 9, 2020. https://www.cnnphilippines.com/
lifestyle/2020/12/9/dogs-dont-really-know-what-you-are-talking-about-study
.html.
33. Humphrey, T., Proops, L., Forman, J., Spooner, R., and McComb, K. "The Role of
Cat Eye Narrowing Movements in Cat–Human Communication." *Scientific Reports*
10 (2020): 16503; Starr, M. "Study Confirms 'Slow Blinks' Really Do Work to Communicate with Your Cat." ScienceAlert, October 8, 2020. https://www.sciencealert
.com/you-can-build-a-rapport-with-your-cat-by-blinking-real-slow.

10. The Faithful Coucal

1. Damore, J. "Google's Ideological Echo Chamber: How Bias Clouds Our Thinking
about Diversity and Inclusion." July 2017.
2. Safari, I., and Goymann, W. "The Evolution of Reversed Sex Roles and Classical
Polyandry: Insights from Coucals and Other Animals." *Ethology* 127 (2021): 1–13.
3. Ibid.
4. Griffiths, P. "Sex Is Real." Aeon. September 21, 2020. https://aeon.co/essays/the
-existence-of-biological-sex-is-no-constraint-on-human-diversity.
5. Ibid.
6. Janssen, D. F. "Know Thy Gender: Etymological Primer." *Archives of Sexual Behavior* 47 (2018): 2149–2154.
7. Haig, D. "The Inexorable Rise of Gender and the Decline of Sex: Social Change in
Academic Titles, 1945–2001." *Archives of Sexual Behavior* 33 (2004): 87–96.
8. Broughton, D. E., Brannigan, R. E., and Omurtag, K. R. "Sex and Gender: You
Should Know the Difference." *Fertility and Sterility* 107 (2017): 1294–1295.
9. Ogle, D. H., and Schanning, K. F. "Usage of 'Sex' and 'Gender.'" *Fisheries* 37 (2012):
271–272.
10. Ibid
11. Rivkis, N. Comment on "If Gender Roles Are Cultural Constructs, How Can There
Be Gender Roles in Animal Species?" Quora, October 28, 2016. https://www.quora

.com/If-gender-roles-are-cultural-constructs-how-can-there-be-gender-roles-in
-animal-species.

12. Sadedin, S. Comment on "What Do Feminists Think of Distinct Gender Roles in Other Species, for Example, in Chickens?" Quora. Accessed August 10, 2017. https://www.quora.com/What-do-feminists-think-of-distinct-gender-roles-in -other-species-for-example-in-chickens.

13. Schwartz, J. "Is Gender Unique to Humans?" *Sapiens*, November 29, 2018. https:// www.sapiens.org/culture/gender-identity-nonhuman-animals/.

14. Ibid.

15. Ah-King, M., and Ahnesjö, I. "The 'Sex Role' Concept: An Overview and Evaluation." *Evolutionary Biology* 40 (2013): 461–470.

16. Lipshutz, S. E., and Rosvall, K. A. "Neuroendocrinology of Sex-Role Reversal." *Integrative and Comparative Biology* 60 (2020): 692–702.

17. Marche, S. "The Unexamined Brutality of the Male Libido."

18. Damore, J. "Google's Ideological Echo Chamber."

19. Ibid.

20. Davis, J. T. M., and Hines, M. "How Large Are Gender Differences in Toy Preferences? A Systematic Review and Meta-analysis of Toy Preference Research." *Archives of Sexual Behavior* 49 (2020): 373–394.

21. Ibid.

22. Pogrebna, G., Oswald, A. J., and Haig, D. "Female Babies and Risk-Aversion: Causal Evidence from Hospital Wards." *Journal of Health Economics* 58 (2018): 10–17.

23. Orr, T. J., and Hayssen, V. "The Female Snark Is Still a Boojum: Looking toward the Future of Studying Female Reproductive Biology." *Integrative and Comparative Biology* 60 (2020): 782–795.

24. https://orwh.od.nih.gov/sex-gender/nih-policy-sex-biological-variable.

25. Eliot, L., and Richardson, S. S. "Sex in Context: Limitations of Animal Studies for Addressing Human Sex/Gender Neurobehavioral Health Disparities." *Journal of Neuroscience* 36, no. 47 (November 2016): 11823–11830.

26. Gururajan, A., Reif, A., Cryan, J. F., and Slattery, D. A. "The Future of Rodent Models in Depression Research." *Nature Reviews Neuroscience* 20 (2019): 686–701.

27. Eliot, L., and Richardson, S. S. "Sex in Context."

28. Rippon, G. "The Trouble with Girls?" *The Psychologist* 29, no. 12 (December 2016): 918–922.

29. Xu, M., Liang, X., Ou, J., Li, H., Luo, Y-j., and Tan, L. H. "Sex Differences in Functional Brain Networks for Language." *Cerebral Cortex* 30 (2020): 1528–1537.

30. Cameron, D. "Evolution, Language and the Battle of the Sexes: A Feminist Linguist Encounters Evolutionary Psychology." *Australian Feminist Studies* 30 (2015): 351–358; Cameron, D. "Sex/Gender, Language and the New Biologism." *Applied Linguistics* 31 (May 2010): 173–192.

31. Ibid.

32. Cameron, D. "Evolution, Language and the Battle of the Sexes."

33. Zuk, M. *Paleofantasy: What Evolution Really Tells Us About Sex, Diet and How We Live.* (2013) New York, W. W. Norton.

34. From: Shahvisi, A. "Nesting Behaviours during Pregnancy: Biological Instinct, or Another Way of Gendering Housework?" *Women's Studies International Forum* 78 (2020): 102329.

35. Ibid.

36. Şahin, Ö., and Yalcinkaya, N. S. "The Gendered Brain: Implications of Exposure to Neuroscience Research for Gender Essentialist Beliefs." *Sex Roles* 84, no. 2 (May 2021): 522–535.

37. Rivkis, ""If Gender Roles Are Cultural Constructs."

11. Protect and Defend

1. Webster, J. P. "The Effect of *Toxoplasma gondii* on Animal Behavior: Playing Cat and Mouse." *Schizophrenia Bulletin* 33 (2007): 752–756.

2. Boillat, M., Hammoudi, P-M., Dogga, S. K., Pagès, S., Goubran, M., Rodriguez, I., and Soldati-Favre, D. "Neuroinflammation-Associated Aspecific Manipulation of Mouse Predator Fear by *Toxoplasma gondii*." *Cell Reports* 30 (2020): 320–334.

3. Servick, K. "Brain Parasite May Strip Away Rodents' Fear of Predators—Not Just of Cats." *Science*, January 14, 2020.

4. Doherty, J-F. "When Fiction Becomes Fact: Exaggerating Host Manipulation by Parasites." *Proceedings of the Royal Society B: Biological Sciences* 287 (2020): 20201081.

5. Fredericksen, M. A., Zhang, Y., Hazen, M. L., Loreto, R. G., Mangold, C. A., Chen, D. Z., and Hughes, D. P. "Three-Dimensional Visualization and a Deep-Learning Model Reveal Complex Fungal Parasite Networks in Behaviorally Manipulated Ants." *Proceedings of the National Academy of Sciences* 114 (2017): 12590–12595.

6. Reiber, C., Shattuck, E. C., Fiore, S., Alperin, P., Davis, V., and Moore, J. "Change in Human Social Behavior in Response to a Common Vaccine." *Annals of Epidemiology* 20, no. 10 (October 2010): 729–733.

7. Ibid.

8. Lopes, P. C. "We Are Not Alone in Trying to Be Alone." *Frontiers in Ecology and Evolution* 8 (2020): 172.

9. Barry, C. "Highly Contagious Coronavirus Variants Powering Another Surge in Europe." *Los Angeles Times*, March 6, 2021.

10. Huffman, M. A. "Self-Medicative Behavior in the African Great Apes: An Evolutionary Perspective into the Origins of Human Traditional Medicine." *BioScience* 51 (August 2001): 651–661.

11. Ibid.

12. Barelli, C., and Huffman, M. A. "Leaf Swallowing and Parasite Expulsion in Khao Yai White-Handed Gibbons (*Hylobates lar*), the First Report in an Asian Ape Species." *American Journal of Primatology* 79 (2017): e22610.

13. Morrogh-Bernard, H. C., Foitová, I., Yeen, Z., Wilkin, P., de Martin, R., Rárová, L., Doležal, K., Nurcahyo, W., and Olšanský, M. "Self-Medication by Orang Utans (*Pongo pygmaeus*) Using Bioactive Properties of *Dracaena cantleyi*." *Scientific Reports* 7 (2017): 16653.

14. de Roode, J. C., Lefèvre, T., and Hunter, M. D. "Self-Medication in Animals." *Science* 340 (2013): 150–151.
15. Villalba, J. J., and Provenza, F. D. "Self-Medication and Homeostatic Behaviour in Herbivores: Learning about the Benefits of Nature's Pharmacy." *Animal* 1 (2007): 1360–1370.
16. Villalba, J. J., Miller, J., Ungar, E. D., Landau, S. Y., and Glendinning, J. "Ruminant Self-Medication against Gastrointestinal Nematodes: Evidence, Mechanism, and Origins." *Parasite* 21 (2014): 31; Costes-Thiré, M., Villalba, J. J., Hoste, H., and Ginane, C. "Increased Intake and Preference for Tannin-Rich Sainfoin (*Onobrychis viciifolia*) Pellets by Both Parasitized and Non-parasitized Lambs after a Period of Conditioning." *Applied Animal Behaviour Science* 203 (2018): 11–18; Lisonbee, L. D., Villalba, J. J., Provenza, F. D., and Hall, J. O. "Tannins and Self-Medication: Implications for Sustainable Parasite Control in Herbivores." *Behavioural Processes* 82 (2009): 184–189.
17. Ibid.
18. Suárez-Rodríguez, M., and Macías Garcia, C. "An Experimental Demonstration That House Finches Add Cigarette Butts in Response to Ectoparasites." *Journal of Avian Biology* 48 (2017): 1316–1321.
19. Uenoyama, R., Miyazaki, T., Hurst, J. L., Beynon, R. J., Adachi, M., Murooka, T., Onoda, I. et al. "The Characteristic Response of Domestic Cats to Plant Iridoids Allows Them to Gain Chemical Defense against Mosquitoes." *Science Advances* 7 (2021): eabd9135.
20. Ibid.
21. Lefèvre, T., Oliver, L., Hunter, M. D., and de Roode, J. C. "Evidence for Trans-generational Medication in Nature." *Ecology Letters* 13 (2010): 1485–1493; Poissonnier, L-A., Lihoreau, M., Gomez-Moracho, T., Dussutour, A., and Buhl, J. "A Theoretical Exploration of Dietary Collective Medication in Social Insects." *Journal of Insect Physiology* 106 (April 2018): 78–87.
22. Simone-Finstrom, M. D., and Spivak, M. "Increased Resin Collection after Parasite Challenge: A Case of Self-Medication in Honey Bees?" *PLoS ONE* 7 (2012): e34601.
23. Spivak, M., Goblirsch, M., and Simone-Finstrom, M. "Social-Medication in Bees: The Line between Individual and Social Regulation." *Current Opinion in Insect Science* 33 (June 2019): 49–55.
24. Frank, E. T., Wehrhahn, M., and Linsenmair, K. E. "Wound Treatment and Selective Help in a Termite-Hunting Ant." *Proceedings of the Royal Society B: Biological Sciences* 285 (2018): 20172457.
25. Greene, A. M., Panyadee, P., Inta, A., and Huffman, M. A. "Asian Elephant Self-Medication as a Source of Ethnoveterinary Knowledge among Karen Mahouts in Northern Thailand." *Journal of Ethnopharmacology* 259 (2020): 112823.
26. Hardy, K., Buckley, S., and Huffman, M. "Doctors, Chefs or Hominin Animals? Non-edible Plants and Neanderthals." *Antiquity* 90 (2016): 1373–1379.
27. McGrew, William M. "In Search of the Last Common Ancestor: New Findings on Wild Chimpanzees." *Philosophical Transactions of the Royal Society B* (2010) 365: 3267–3276.
28. Sherman, P. W., and Billing, J. "Darwinian Gastronomy: Why We Use Spices: Spices Taste Good Because They Are Good for Us." *BioScience* (1999) 49: 453–463;

Billing, J., and Sherman, P. W. "Antimicrobial Functions of Spices: Why Some Like It Hot." *Quarterly Review of Biology* (1998) 73: 3–49.

29. Ibid.

30. Bromham, L., Skeels, A., Schneemann, H., Dinnage, R., and Hua, X. "There Is Little Evidence That Spicy Food in Hot Countries Is an Adaptation to Reducing Infection Risk." *Nature Human Behaviour* 5 (2021): 878–891; Bromhan, L. "Why Do Hot Countries Have Spicy Food?" *Behind the Paper* (blog), SocialSciences.Nature, February 4, 2021.

31. Bromhan, L. "Why Do Hot Countries Have Spicy Food?"

32. Ibid.

33. Makin, D. F., Kotler, B. P., Brown, J. S., Garrido, M., and Menezes, J. F. S. "The Enemy Within: How Does a Bacterium Inhibit the Foraging Aptitude and Risk Management Behavior of Allenby's Gerbils?" *American Naturalist* 196 (2020): 717–729.

34. Ibid.

35. Mitoh, S., and Yusa, Y. "Extreme Autotomy and Whole-Body Regeneration in Photosynthetic Sea Slugs." *Current Biology* 31 (2021): R233–234; Coyne, J. "Sea Slug Regrows Entire Body from Just the Decapitated Head, or 'Autotomy with Kleptoplasty.'" *Why Evolution Is True* (blog), March 9, 2021. https://whyevolutionistrue.com/2021/03/09/sea-slug-regrows-entire-body-from-just-the-decapitated-head-or-autotomy-with-kleptoplasty/.

Bibliography

Introduction

Barlow, Fiona Kate. "Nature vs. Nurture Is Nonsense: On the Necessity of an Integrated Genetic, Social, Developmental, and Personality Psychology." *Australian Journal of Psychology* 71, no. 1 (March 2019): 68–79. https://doi.org/10.1111/ajpy.12240.

Christensen, K. D., T. E. Jayaratne, J. S. Roberts, S. L. R. Kardia, and E. M. Petty. "Understandings of Basic Genetics in the United States: Results from a National Survey of Black and White Men and Women." *Public Health Genomics* 13, no. 7–8 (December 2010): 467–476. https://doi.org/10.1159/000293287.

Clarkin, Patrick F. "We Are Not Hard-Wired." *This View of Life* (blog), August 22, 2017. https://evolution-institute.org/article/we-are-not-hard-wired/?source=tvol (site discontinued).

Comfort, Nathaniel. "Genetic Determinism Redux." *Nature* 561 (September 27, 2018): 461–463.

———. "Lies, Damned Lies, and GWAS." *Genotopia* (blog), October 5, 2018. https://genotopia.scienceblog.com/506/lies-damned-lies-and-gwas/.

Condit, Celeste M. "How the Public Understands Genetics: Non-Deterministic and Non-Discriminatory Interpretations of the 'Blueprint' Metaphor." *Public Understanding of Science* 8, no. 3 (July 1999): 169–180. https://doi.org/10.1088/0963-6625/8/3/302.

———. "Laypeople Are Strategic Essentialists, Not Genetic Essentialists." In "Looking for the Psychosocial Impacts of Genomic Information." Special report, *Hastings Center Report* 49, no. 3 (May 2019): S27–37. https://doi.org/10.1002/hast.1014.

Condit, Celeste M., Benjamin R. Bates, Ryan Galloway, Sonja Brown Givens, Caroline K. Haynie, John W. Jordan, Gordon Stables, and Hollis Marshall West. "Recipes or Blueprints for Our Genes? How Contexts Selectively Activate the Multiple Meanings of Metaphors." *Quarterly Journal of Speech* 88, no. 3 (August 2002): 303–325. https://doi.org/10.1080/00335630209384379.

Fall, Tove, Ralf Kuja-Halkola, Keith Dobney, Carri Westgarth, and Patrik K. E. Magnusson. "Evidence of Large Genetic Influences on Dog Ownership in the Swedish Twin Registry Has Implications for Understanding Domestication and Health Associations." *Scientific Reports* 9, no. 1 (December 2019): 7554. https://doi.org/10.1038/ s41598-019-44083-9.

Freestone, Jamie Milton. "Human Exceptionalism with a Human Face." *Areo*, April 19, 2019. https://areomagazine.com/2019/04/19/human-exceptionalism-with-a -human-face/.

Heine, S. J., I. Dar-Nimrod, B. Y. Cheung, and T. Proulx. "Essentially Biased: Why People Are Fatalistic about Genes." *Advances in Experimental Social Psychology* 55 (2017): 137–192. https://doi.org/10.1016/bs.aesp.2016.10.003.

Heine, Steven J., Benjamin Y. Cheung, and Anita Schmalor. "Making Sense of Genetics: The Problem of Essentialism." In "Looking for the Psychosocial Impacts of Genomic Information." Special report, *Hastings Center Report* 49, no. 3 (May 2019): S19–26. https://doi.org/10.1002/hast.1013.

Hill, W. David, Neil M. Davies, Stuart J. Ritchie, Nathan G. Skene, Julien Bryois, Steven Bell, Emanuele Di Angelantonio, et al. "Genome-Wide Analysis Identifies Molecular Systems and 149 Genetic Loci Associated with Income." *Nature Communications* 10, no. 1 (December 2019): 5741. https://doi.org/10.1038/s41467-019-13585-5.

Junger, Sebastian. "Our Politics Are in Our DNA." *Washington Post*, July 7, 2019.

Kaufman, Scott Barry. "There Is No Nature-Nurture War." *Beautiful Minds* (blog), *Scientific American*, January 18, 2019. https://blogs.scientificamerican.com/beautiful -minds/there-is-no-nature-nurture-war/.

Landau, Barbara. "The Importance of the Nativist-Empiricist Debate: Thinking About Primitives Without Primitive Thinking." *Child Development Perspectives* 3, no. 2 (August 2009): 88–90. https://doi.org/10.1111/j.1750-8606.2009.00082.x.

Lee, James J., Robbee Wedow, Aysu Okbay, Edward Kong, Omeed Maghzian, Meghan Zacher, Tuan Anh Nguyen-Viet, et al. "Gene Discovery and Polygenic Prediction from a Genome-Wide Association Study of Educational Attainment in 1.1 Million Individuals." *Nature Genetics* 50, no. 8 (July 2018): 1112–1121. https://doi.org/10 .1038/s41588-018-0147-3.

Letsinger, Ayland C., Jorge Z. Granados, Sarah E. Little, and J. Timothy Lightfoot. "Alleles Associated with Physical Activity Levels Are Estimated to Be Older than Anatomically Modern Humans." *PLoS ONE* 14, no. 4 (April 29, 2019): e0216155. https://doi.org/10.1371/journal.pone.0216155.

Lewkowicz, David J. "The Biological Implausibility of the Nature-Nurture Dichotomy and What It Means for the Study of Infancy." *Infancy* 16, no. 4 (July 2011): 331–367. https://doi.org/10.1111/j.1532-7078.2011.00079.x.

Macdonald, Helen. "What Animals Taught Me about Being Human." *New York Times Magazine*, May 16, 2017. https://www.nytimes.com/2017/05/16/magazine/what -animals-taught-me-about-being-human.html.

Marche, S. "The Unexamined Brutality of the Male Libido." *New York Times*, November 25, 2017.

Paul, Diane B. "The Nine Lives of Discredited Data." *The Sciences* 27, no. 3 (June 1987): 26–30.

Paul, Diane B., and J. P. Brosco. *The PKU Paradox.* (2013) Baltimore, NY, Johns Hopkins University Press.

Perrault, Sarah Tinker, and Meaghan O'Keefe. "New Metaphors for New Understandings of Genomes." *Perspectives in Biology and Medicine* 62, no. 1 (Winter 2019): 1–19. https://doi.org/10.1353/pbm.2019.0000.

Plomin, Robert. "In the Nature-Nurture War, Nature Wins." *Observations* (blog), *Scientific American*, December 14, 2018. https://blogs.scientificamerican.com/observations/in-the-nature-nurture-war-nature-wins/.

Plomin, Robert, John C. DeFries, Valerie S. Knopik, and Jenae M. Neiderhiser. "Top 10 Replicated Findings from Behavioral Genetics." *Perspectives on Psychological Science* 11, no. 1 (January 2016): 3–23. https://doi.org/10.1177/1745691615617439.

Reimann, Martin, Oliver Schilke, and Karen S. Cook. "Trust Is Heritable, Whereas Distrust Is Not." *Proceedings of the National Academy of Sciences* 114, no. 27 (July 3, 2017): 7007–7012. https://doi.org/10.1073/pnas.1617132114.

Slijper, E. J. "Biologic-Anatomical Investigations on the Bipedal Gait and Upright Posture in Mammals, with Special Reference to a Little Goat, Born without Forelegs." *Proceedings of the Koninklijke Nederlandse Akademie Van Wetenschappen* 45 (1942): 288–295, 407–415.

Spelke, Elizabeth S., and Katherine D. Kinzler. "Innateness, Learning, and Rationality." *Child Development Perspectives* 3, no. 2 (August 2009): 96–98. https://doi.org/10.1111/j.1750-8606.2009.00085.x.

Spencer, John P., Mark S. Blumberg, Bob McMurray, Scott R. Robinson, Larissa K. Samuelson, and J. Bruce Tomblin. "Short Arms and Talking Eggs: Why We Should No Longer Abide the Nativist-Empiricist Debate." *Child Development Perspectives* 3, no. 2 (August 2009): 79–87. https://doi.org/10.1111/j.1750-8606.2009.00081.x.

Turkheimer, Eric. "Cochran on Zimmer, and Correcting an Old Misimpression." *Genetics & Human Agency* (blog), July 5, 2018. https://www.geneticshumanagency.org/gha/cochran-on-zimmer-and-correcting-an-old-misimpression/.

———. "More on Murray: What Is Biological Determinism?" *Genetics & Human Agency* (blog), January 29, 2020. https://www.geneticshumanagency.org/gha/more-on-murray-what-is-biological-determinism/.

———. "The Blueprint Metaphor." *Genetics & Human Agency* (blog), October 30, 2018. https://www.geneticshumanagency.org/gha/the-blueprint-metaphor/.

———. "The Gloomy Prospect Then and Now." *Genetics & Human Agency* (blog), November 20, 2018. https://www.geneticshumanagency.org/gha/the-gloomy-prospect-then-and-now/.

———. "The Social Science Blues." *Hastings Center Report* 49, no. 3 (June 2019): 45–47. https://doi.org/10.1002/hast.1008.

———. "Three Laws of Behavior Genetics and What They Mean." *Current Directions in Psychological Science* 9, no. 5 (October 2000): 160–164. https://doi.org/10.1111/1467-8721.00084.

———. "Weak Genetic Explanation 20 Years Later: Reply to Plomin et al. (2016)." *Perspectives on Psychological Science* 11, no. 1 (2016): 24–28. https://doi.org/10.1177/1745691615617442.

University of Iowa. "Nature? Nurture? Child Development Scientists Say Neither." *ScienceDaily*, July 21, 2009. https://www.sciencedaily.com/releases/2009/07/090720163723.htm.

Waal, Frans de. "Closer to Beast than Angel." *Los Angeles Review of Books*, May 14, 2018. https://lareviewofbooks.org/article/closer-to-beast-than-angel/.

West-Eberhard, Mary Jane. "Developmental Pasticity and the Origin of Species Differences." In *Systematics and the Origin of Species: On Ernst Mayr's 100th Anniversary*, edited by Jody Hey, Walter M. Fitch, and Francisco J. Ayala, 69–89. (2005) Washington, DC, The National Academies Press.

Will, George F. "Is the Individual Obsolete?" Minneapolis *Star Tribune*, June 16, 2019. www.startribune.com/is-the-individual-obsolete/511325222/.

1. Narwhals and the Dead Man

Abramson, Charles I., and Paco Calvo. "General Issues in the Cognitive Analysis of Plant Learning and Intelligence." In *Memory and Learning in Plants*, edited by Frantisek Baluska, Monica Gagliano, and Guenther Witzany, 35–49. Signaling and Communication in Plants series. (2018) Cham, Switzerland, Springer. https://doi.org/10.1007/978-3-319-75596-0_3.

Angier, Natalie. "When 'What Animals Do' Doesn't Seem to Cover It." *New York Times*, July 20, 2009. https://www.nytimes.com/2009/07/21/science/21angier.html.

Baum, William M. "What Counts as Behavior? The Molar Multiscale View." *The Behavior Analyst* 36, no. 2 (2013): 283–293. https://doi.org/10.1007/BF03392315.

Bergner, Raymond M. "What Is Behavior? And So What?" *New Ideas in Psychology* 29, no. 2 (August 2011): 147–155. https://doi.org/10.1016/j.newideapsych.2010.08.001.

———. "What Is Behavior? And Why Is It Not Reducible to Biological States of Affairs?" *Journal of Theoretical and Philosophical Psychology* 36, no. 1 (2016): 41–55. https://doi.org/10.1037/teo0000026.

Bertossa, Rinaldo C. "Morphology and Behaviour: Functional Links in Development and Evolution." *Philosophical Transactions of the Royal Society B: Biological Sciences* 366, no. 1574 (July 27, 2011): 2056–2068. https://doi.org/10.1098/rstb.2011.0035.

Böhm, Jennifer, Sönke Scherzer, Elzbieta Krol, Ines Kreuzer, Katharina von Meyer, Christian Lorey, Thomas D. Mueller, et al. "The Venus Flytrap *Dionaea muscipula* Counts Prey-Induced Action Potentials to Induce Sodium Uptake." *Current Biology* 26, no. 3 (February 8, 2016): 286–295. https://doi.org/10.1016/j.cub.2015.11.057.

Buston, Peter. "Size and Growth Modification in Clownfish." *Nature* 424, no. 6945 (July 10, 2003): 145–146. https://doi.org/10.1038/424145a.

Buston, Peter M., and Michael A. Cant. "A New Perspective on Size Hierarchies in Nature: Patterns, Causes, and Consequences." *Oecologia* 149, no. 2, June 23, 2006: 362–372. https://doi.org/10.1007/s00442-006-0442-z.

Candea, Matei. "Behaviour as a Thing." *Interdisciplinary Science Reviews* 44, no. 1 (January 2, 2019): 1–11. https://doi.org/10.1080/03080188.2018.1561064.

Critchfield, Thomas S. "Requiem for the Dead Man Test?" *The Behavior Analyst* 40, no. 2 (November 2017): 539–548. https://doi.org/10.1007/s40614-016-0082-5.

Critchfield, T. S., and Shue, E. Z. H. "The Dead Man Test: a Preliminary Experimental Analysis." *Behavior Analysis in Practice* 11 (2018): 381–384.

Cvrčková, Fatima, Viktor Žárský, and Anton Markoš. "Plant Studies May Lead Us to

Rethink the Concept of Behavior." *Frontiers in Psychology* 7 (April 28, 2016): 622. https://doi.org/10.3389/fpsyg.2016.00622.

Dawkins, R. *The Selfish Gene*. (1976) Oxford, Oxford University Press.

Giaimo, Cara. "Praying Mantises: More Deadly than We Knew." *New York Times*, May 14, 2020. https://www.nytimes.com/2020/05/14/science/praying-mantis-strike.html.

———. "Watch a Flower That Seems to Remember When Pollinators Will Come Calling." *New York Times*, April 20, 2019. https://www.nytimes.com/2019/04/20/science/plants-moving-memory.html.

Henriques, Gregg, and Joseph Michalski. "Defining Behavior and Its Relationship to the Science of Psychology." *Integrative Psychological and Behavioral Science* (2020) 54: 328–353.

Hou, Chia-Yi. "Botanists Say Plants Are Not Conscious." *The Scientist*, July 5, 2019. https://www.the-scientist.com/news-opinion/botanists-say-plants-are-not-conscious-66101.

Huchard, Elise, Sinead English, Matt B. V. Bell, Nathan Thavarajah, and Tim Clutton-Brock. "Competitive Growth in a Cooperative Mammal." *Nature* 533, no. 7604 (May 2016): 532–534. https://doi.org/10.1038/nature17986.

Klein, Alice. "Bees Force Plants to Flower Early by Cutting Holes in Their Leaves." *New Scientist*, May 21, 2020. https://www.newscientist.com/article/2244009-bees-force-plants-to-flower-early-by-cutting-holes-in-their-leaves/.

Levitis, Daniel A., William Z. Lidicker Jr., and Glenn Freund. "Behavioural Biologists Do Not Agree on What Constitutes Behaviour." *Animal Behaviour* 78, no. 1 (July 2009): 103–110. https://doi.org/10.1016/j.anbehav.2009.03.018.

Duijn, Marc van, Fred Keijzer, and Daan Franken. "Principles of Minimal Cognition: Casting Cognition as Sensorimotor Coordination." *Adaptive Behavior* 14, no. 2 (June 2006): 157–170. https://doi.org/10.1177/105971230601400207.

Pashalidou, Foteini G., Harriet Lambert, Thomas Peybernes, Mark C. Mescher, and Consuelo M. De Moraes. "Bumble Bees Damage Plant Leaves and Accelerate Flower Production When Pollen Is Scarce." *Science* 368, no. 6493 (May 22, 2020): 881–884. https://doi.org/10.1126/science.aay0496.

Pennisi, Elizabeth. "Melting Sea Ice Is Stressing Out Narwhals." *Science*, December 7, 2017. https://doi.org/10.1126/science.aar6991.

Reed, Cymone, Rebecca Branconi, John Majoris, Cara Johnson, and Peter Buston. "Competitive Growth in a Social Fish." *Biology Letters* 15, no. 2 (February 2019): 20180737. https://doi.org/10.1098/rsbl.2018.0737.

Reid, C. R., H. Macdonald, R. P. Mann, J. A. R. Marshall, T. Latty, and S. Garnier. "Decision-Making without a Brain: How an Amoeboid Organism Solves the Two-Armed Bandit." *Journal of the Royal Society Interface* 13 (2016) http://dx.doi.org/10.1098/rsif.2016.0030.

Rossoni, Sergio, and Jeremy E. Niven. "Prey Speed Influences the Speed and Structure of the Raptorial Strike of a 'Sit-and-Wait' Predator." *Biology Letters* 16, no. 5 (May 2020): 20200098. https://doi.org/10.1098/rsbl.2020.0098.

Taiz, Lincoln, Daniel Alkon, Andreas Draguhn, Angus Murphy, Michael Blatt, Chris Hawes, Gerhard Thiel, and David G. Robinson. "Plants neither Possess nor Require

Consciousness." *Trends in Plant Science* 24, no. 8 (August 2019): 677–687. https://doi .org/10.1016/j.tplants.2019.05.008.

Tinbergen, N. *The Study of Instinct.* (1951) Oxford, Oxford University Press.

Uher, Jana. "What Is Behaviour? And (When) Is Language Behaviour? A Metatheoretical Definition." *Journal for the Theory of Social Behaviour* 46, no. 4 (December 2016): 475–501. https://doi.org/10.1111/jtsb.12104.

Williams, T. M., S. B. Blackwell , B. Richter, M-H. S. Sinding, M. P. Heide-Jørgensen. "Paradoxical Escape Responses by Narwhals (*Monodon monoceros*)." *Science* 358 (2017): 1328–1331.

2. Snakes, Spiders, Bees, and Princesses

Bateson, Patrick. "The Active Role of Behaviour in Evolution." *Biology and Philosophy* 19, no. 2 (March 2004): 283–298. https://doi.org/10.1023/B:BIPH.0000024468.12161.83.

Bostanchi, Hamid, Steven C. Anderson, Haji Gholi Kami, and Theodore J. Papenfuss. "A New Species of *Pseudocerastes* with Elaborate Tail Ornamentation from Western Iran (Squamata: Viperidae)." *Proceedings of the California Academy of Sciences*, Fourth Series, 57, no. 14 (September 15, 2006): 443–450.

Bostwick, Kimberly S. "Display Behaviors, Mechanical Sounds, and Evolutionary Relationships of the Club-Winged Manakin (*Machaeropterus deliciosus*)." *Auk* 117 (2000): 465–478.

Bostwick, Kimberly S., M .L. Riccio, and J. M. Humphries. "Massive, Solidified Bone in the Wing of a Volant Courting Bird." *Biology Letters* 8 (2012): 760–763.

Burghardt, Gordon M. "Darwin's Legacy to Comparative Psychology and Ethology." *American Psychologist* 64, no. 2 (February 2009): 102–110. https://doi.org/10.1037/ a0013385.

———. "Ground Rules for Dealing with Anthropomorphism." *Nature* 430 (July 1, 2004): 15.

Burghardt, Gordon M., and Harold A. Herzog Jr. "Beyond Conspecifics: Is Brer Rabbit Our Brother?" *BioScience* 30, no. 11 (1980): 763–768.

Cesario, Joseph, David J. Johnson, and Heather L. Eisthen. "Your Brain Is Not an Onion with a Tiny Reptile Inside." *Current Directions in Psychological Science* 29, no. 3 (2020): 255–260. https://doi.org/10.1177/0963721420917687.

Corning, Peter A. "Evolution 'on Purpose': How Behaviour Has Shaped the Evolutionary Process." *Biological Journal of the Linnean Society* 112, no. 2 (June 2014): 242–260. https://doi.org/10.1111/bij.12061.

Darwin, C. *The Expression of the Emotions in Man and Animals.* (1872) London, John Murray.

de Waal, F. *Mama's Last Hug: Animal Emotions and What They Tell Us about Ourselves.* (2019) New York, W. W. Norton.

Emery, N. J. "Evolution of Learning and Cognition." In J. Call, G. M. Burghardt, I. M. Pepperberg, C. T. Snowdon, and T. Zentall (Eds.), *APA Handbook of Comparative Psychology: Basic Concepts, Methods, Neural Substrate, and Behavior* (2017): 237–255. American Psychological Association. https://doi.org/10.1037/0000011-012.

Farah, Troy. "Meet the Snake That Hunts Birds with a Spider on Its Tail." *Discover*, April

16, 2019. https://www.discovermagazine.com/planet-earth/meet-the-snake-that
-hunts-birds-with-a-spider-on-its-tail.

Greene, Harry W. "Evolutionary Scenarios and Primate Natural History." *The American Naturalist* 190, no. S1 (August 2017): S69–86. https://doi.org/10.1086/692830.

———. "Homology and Behavioral Repertoires." In *Homology: The Hierarchical Basis of Comparative Biology*, edited by Brian K. Hall, 369–391. (1994) San Diego, CA: Academic Press.

———. "Natural History and Behavioural Homology." In *Homology*, 173–188. Novartis Foundation Symposium 222. (1999) Chichester, UK: Wiley.

Hall, Brian K. "Homology, Homoplasy, Novelty, and Behavior." *Developmental Psychobiology* 55, no. 1 (January 2013): 4–12. https://doi.org/10.1002/dev.21039.

Jabr, Ferris. "The Evolution of Emotion: Charles Darwin's Little-Known Psychology Experiment." *Observations* (blog), *Scientific American*, May 24, 2010. https://blogs .scientificamerican.com/observations/the-evolution-of-emotion-charles-darwins -little-known-psychology-experiment/?print=true.

Klopfer, Peter H. "Behavior." Review of *Ethology of Mammals*, by R. F. Ewer. *Science*, n.s., 165, no. 3896 (August 29, 1969): 887. https://doi.org/10.1126/science.165.3896.887.

———. "Does Behavior Evolve?" *Annals of the New York Academy of Sciences* 223, no. 1 (December 1973): 113–119. https://doi.org/10.1111/j.1749-6632.1973.tb41425.x.

———. "Instincts and Chromosomes: What Is an 'Innate' Act?" *The American Naturalist* 103, no. 933 (September 1969): 556–560. https://doi.org/10.1086/282624.

———. Review of *Animal Behavior: An Evolutionary Approach*, by John Alcock, and *Defence in Animals*, by M. Edmunds. *American Scientist* 63, no. 5 (October 1975): 578–579.

———. "Still Largely Where Darwin Left Us." *PsycCRITIQUES* 20, no. 5 (May 1975): 406–407. https://doi.org/10.1037/0013339.

MacLean, P. D. "Man and His Animal Brains." *Modern Medicine* 32 (1964): 95–106.

Marshall, Michael. "Rats Will Help Others in Distress, But They Can Be Influenced Not To." *New Scientist*, July 8, 2020. https://www.newscientist.com/article/2248324 -rats-will-help-others-in-distress-but-they-can-be-influenced-not-to/.

Queiroz, Alan de, and Peter H. Wimberger. "The Usefulness of Behavior for Phylogeny Estimation: Levels of Homoplasy in Behavioral and Morphological Characters." *Evolution* 47, no. 1 (1993): 44–60.

Reid, Chris R., Hannelore MacDonald, Richard P. Mann, James A. R. Marshall, Tanya Latty, and Simon Garnier. "Decision-Making without a Brain: How an Amoeboid Organism Solves the Two-Armed Bandit." *Journal of the Royal Society Interface* 13, no. 119 (June 2016): 20160030. https://doi.org/10.1098/rsif.2016.0030.

Sagan, Carl. *The Dragons of Eden*. (1981) New York, Ballantine Books.

Strassmann, Joan E. "Tribute to Tinbergen: The Place of Animal Behavior in Biology." *Ethology* 120, no. 2 (2014): 123–126. https://doi.org/10.1111/eth.12192.

Vicedo, Marga. "The 'Disadapted' Animal: Niko Tinbergen on Human Nature and the Human Predicament." *Journal of the History of Biology* 51, no. 2 (June 2018): 191–221. https://doi.org/10.1007/s10739-017-9485-8.

Wenzel, John W. "Behavioral Homology and Phylogeny." *Annual Review of Ecology and Systematics* 23 (November 1992): 361–381.

3. Clean-Minded Bees and Courtship Genes

Bastock, Margaret. "A Gene Mutation Which Changes a Behavior Pattern." *Evolution* 10 (December 1956): 421–439.

Bell, Alison M., and Gene E. Robinson. "Behavior and the Dynamic Genome." *Science* 332, no. 6034 (June 3, 2011): 1161–1162. https://doi.org/10.1126/science.1203295.

Belluck, Pam. "Many Genes Influence Same-Sex Sexuality, Not a Single 'Gay Gene.'" *New York Times*, August 29, 2019. https://www.nytimes.com/2019/08/29/science/gay-gene-sex.html.

Benton, Michael J. "Studying Function and Behavior in the Fossil Record." *PLoS Biology* 8, no. 3 (March 2010): e1000321. https://doi.org/10.1371/journal.pbio.1000321.

Boake, Christine R. B., Stevan J. Arnold, Felix Breden, Lisa M. Meffert, Michael G. Ritchie, Barbara J. Taylor, Jason B. Wolf, and Allen J. Moore. "Genetic Tools for Studying Adaptation and the Evolution of Behavior." *American Naturalist* 160, no. S6 (December 2002): S143–159. https://doi.org/10.1086/342902.

Braudt, David B. "Sociogenomics in the 21st Century: An Introduction to the History and Potential of Genetically Informed Social Science." *Sociology Compass* 12, no. 10 (October 2018): e12626. https://doi.org/10.1111/soc4.12626.

Chabris, Christopher F., James J. Lee, David Cesarini, Daniel J. Benjamin, and David I. Laibson. "The Fourth Law of Behavior Genetics." *Current Directions in Psychological Science* 24, no. 4 (August 2015): 304–312. https://doi.org/10.1177/0963721415580430.

Cobb, Matthew. "A Gene Mutation Which Changed Animal Behaviour: Margaret Bastock and the Yellow Fly." *Animal Behaviour* 74, no. 2 (August 2007): 163–169. https://doi.org/10.1016/j.anbehav.2007.05.002.

Dilger, William C. "The Behavior of Lovebirds." *Scientific American* 206, no. 1 (January 1962): 88–99.

Donovan, Brian M. "Looking Backwards to Move Biology Education toward Its Humanitarian Potential: A Review of *Darwinism, Democracy, and Race*." Review of *Darwinism, Democracy, and Race*, by John P. Jackson Jr. and David J. Depew. *Science Education* 102, no. 6 (2018): 1399–1404. https://doi.org/10.1002/sce.21480.

———. "Putting Humanity Back into the Teaching of Human Biology." *Studies in History and Philosophy of Biological and Biomedical Sciences* 52 (August 2015): 65–75. https://doi.org/10.1016/j.shpsc.2015.01.011.

Donovan, Brian M., Rob Semmens, Phillip Keck, Elizabeth Brimhall, K. C. Busch, Monica Weindling, Alex Duncan, et al. "Toward a More Humane Genetics Education: Learning about the Social and Quantitative Complexities of Human Genetic Variation Research Could Reduce Racial Bias in Adolescent and Adult Populations." *Science Education* 103, no. 3 (2019): 529–560. https://doi.org/10.1002/sce.21506.

Donovan, Brian M., Molly A. M. Stuhlsatz, Daniel C. Edelson, and Zoë E. Buck Bracey. "Gendered Genetics: How Reading about the Genetic Basis of Sex Differences in Biology Textbooks Could Affect Beliefs Associated with Science Gender Disparities." *Science Education* 103, no. 4 (2019): 719–749. https://doi.org/10.1002/sce.21502.

Ebstein, Richard P., Salomon Israel, Soo Hong Chew, Songfa Zhong, and Ariel Knafo. "Genetics of Human Social Behavior." *Neuron* 65, no. 6 (March 2010): 831–844. https://doi.org/10.1016/j.neuron.2010.02.020.

Kendler, Kenneth S., and Ralph J. Greenspan. "The Nature of Genetic Influences on Behavior: Lessons from 'Simpler' Organisms." *American Journal of Psychiatry* 163, no. 10 (2006): 1683–1694.

Meffert, Lisa M., Sara K. Hicks, and Jennifer L. Regan. "Nonadditive Genetic Effects in Animal Behavior." *American Naturalist* 160, no. S6 (December 2002): S198–213. https://doi.org/10.1086/342896.

Pan, Yufeng, and Bruce S. Baker. "Genetic Identification and Separation of Innate and Experience-Dependent Courtship Behaviors in *Drosophila*." *Cell* 156, no. 1–2 (January 2014): 236–248. https://doi.org/10.1016/j.cell.2013.11.041.

Polderman, T., B. Benyamin, C. de Leeuw, P. F. Sullivan, A. van Bochoven, P. M. Visscher, and D. Posthum. "Meta-analysis of the Heritability of Human Traits Based on Fifty Years of Twin Studies." *Nature Genetics* 47 (2015): 702–709.

Rothenbuhler, W. C. "Behavior Genetics of Nest Cleaning in Honey Bees. IV. Responses of F1 and Backcross Generations to Disease-Killed Brood." *American Zoologist* 4 (1964): 111–123.

Scannapieco, Alejandra C., Silvia B. Lanzavecchia, María A. Parreño, María C. Liendo, Jorge L. Cladera, Marla Spivak, and María A. Palacio. "Individual Precocity, Temporal Persistence, and Task-Specialization of Hygienic Bees from Selected Colonies of *Apis mellifera*." *Journal of Apicultural Science* 60, no. 1 (2016): 49–60. https://doi.org/10.1515/jas-2016-0006.

Shpigler, Hagai Y., Michael C. Saul, Frida Corona, Lindsey Block, Amy Cash Ahmed, Sihai D. Zhao, and Gene E. Robinson. "Deep Evolutionary Conservation of Autism-Related Genes." *Proceedings of the National Academy of Sciences* 114, no. 36 (September 5, 2017): 9653–9658. https://doi.org/10.1073/pnas.1708127114.

Wahlsten, Douglas. "A Contemporary View of Genes and Behavior: Complex Systems and Interactions." In *Advances in Child Development and Behavior*, edited by Richard M. Lerner and Janette B. Benson, 44:285–306. Elsevier, 2013. https://doi.org/10.1016/B978-0-12-397947-6.00010-6.

———. "Single-Gene Influences on Brain and Behavior." *Annual Review of Psychology* 50 (February 1999): 599–624. https://doi.org/10.1146/annurev.psych.50.1.599.

Williams, Terrie M., Susanna B. Blackwell, Beau Richter, Mikkel-Holger S. Sinding, and Mads Peter Heide-Jørgensen. "Paradoxical Escape Responses by Narwhals (*Monodon monoceros*)." *Science* 358, no. 6368 (December 8, 2017): 1328–1331. https://doi.org/10.1126/science.aao2740.

York, Ryan A. "Assessing the Genetic Landscape of Animal Behavior." *Genetics* 209, no. 1 (May 2018): 223–232. https://doi.org/10.1534/genetics.118.300712.

Zimmer, Carl. *She Has Her Mother's Laugh: The Powers, Perversions, and Potential of Heredity.* (2018) New York, Penguin Random House.

4. Raised by Wolves—Would It Really Be So Bad?

Achterberg, Puck. "The Reason Why Cats Don't Behave, While Dogs Do." *Blue Skies & Happy Clouds* (blog), July 20, 2020. https://medium.com/blue-skies-happy-clouds/the-reason-why-cats-dont-behave-while-dogs-do-b59446198c70.

Ahmad, Hafiz Ishfaq, Muhammad Jamil Ahmad, Farwa Jabbir, Sunny Ahmar, Nisar

Ahmad, Abdelmotaleb A. Elokil, and Jinping Chen. "The Domestication Makeup: Evolution, Survival, and Challenges." *Frontiers in Ecology and Evolution* 8 (May 8, 2020): 103. https://doi.org/10.3389/fevo.2020.00103.

Alpi, Kristine M., and Barbara L. Sherman. "The Well-Behaved Dog." *Library Journal*, November 1, 2008.

Anderson, Eugene N. Review of *The First Domestication: How Wolves and Humans Coevolved*, by Raymond Pierotti and Brandy R. Fogg. *Ethnobiology Letters* 9, no. 2 (October 5, 2018): 247–249. https://doi.org/10.14237/ebl.9.2.2018.1379.

Arizona State University. "Yes, Your Dog Wants to Rescue You: Pet Dogs Will Try to Save Their Distressed Human, as Long as They Know How." *ScienceDaily*, May 29, 2020. https://www.sciencedaily.com/releases/2020/05/200529150706.htm.

Arnott, Elizabeth R., Lincoln Peek, Jonathan B. Early, Annie Y. H. Pan, Bianca Haase, Tracy Chew, Paul D. McGreevy, and Claire M. Wade. "Strong Selection for Behavioural Resilience in Australian Stock Working Dogs Identified by Selective Sweep Analysis." *Canine Genetics and Epidemiology* 2, no. 6 (December 2015). https://doi.org/10.1186/s40575-015-0017-6.

Asch, Barbara van, Ai-bing Zhang, Mattias C. R. Oskarsson, Cornelya F. C. Klütsch, António Amorim, and Peter Savolainen. "Pre-Columbian Origins of Native American Dog Breeds, with Only Limited Replacement by European Dogs, Confirmed by mtDNA Analysis." *Proceedings of the Royal Society B: Biological Sciences* 280, no. 1766 (2013): 20131142. https://doi.org/10.1098/rspb.2013.1142.

Bekoff, Mark. "The First Domestication: How Wolves and Humans Coevolved." *Psychology Today*, December 11, 2017. https://www.psychologytoday.com/blog/animal-emotions/201712/the-first-domestication-how-wolves-and-humans-coevolved.

Botigué, Laura R., Shiya Song, Amelie Scheu, Shyamalika Gopalan, Amanda L. Pendleton, Matthew Oetjens, Angela M. Taravella, et al. "Ancient European Dog Genomes Reveal Continuity since the Early Neolithic." *Nature Communications* 8 (July 2017): 16082. https://doi.org/10.1038/ncomms16082.

Boyko, Adam R., Pascale Quignon, Lin Li, Jeffrey J. Schoenebeck, Jeremiah D. Degenhardt, Kirk E. Lohmueller, Keyan Zhao, et al. "A Simple Genetic Architecture Underlies Morphological Variation in Dogs." *PLoS Biology* 8, no. 8 (August 2010): e1000451. https://doi.org/10.1371/journal.pbio.1000451.

Chijiiwa, Hitomi, Hika Kuroshima, Yusuke Hori, James R. Anderson, and Kazuo Fujita. "Dogs Avoid People Who Behave Negatively to Their Owner: Third-Party Affective Evaluation." *Animal Behaviour* 106 (August 2015): 123–127. https://doi.org/10.1016/j.anbehav.2015.05.018.

Darwin, C. *The Variation of Plants and Animals Under Domestication.* (1868) London, John Murray.

Dodman, N. H., E. K. Karlsson, A. Moon-Fanelli, M. Galdzicka, M. Perloski, L. Shuster, K. Lindblad-Toh, and E. I. Ginns. "A Canine Chromosome 7 Locus Confers Compulsive Disorder Susceptibility." *Molecular Psychiatry* 15, no. 1 (January 2010): 8–10. https://doi.org/10.1038/mp.2009.111.

"Do Dogs Actually Love Us?" Quora. Accessed August 27, 2017. https://www.quora.com/Do-dogs-actually-love-us.

Drake, Abby Grace, and Christian Peter Klingenberg. "Large-Scale Diversification of

Skull Shape in Domestic Dogs: Disparity and Modularity." *American Naturalist* 175, no. 3 (March 2010): 289–301. https://doi.org/10.1086/650372.

Driscoll, Carlos A., David W. Macdonald, and Stephen J. O'Brien. "From Wild Animals to Domestic Pets, an Evolutionary View of Domestication." *Proceedings of the National Academy of Sciences* 106, Supplement 1 (June 16, 2009): 9971–9978. https://doi.org/10.1073/pnas.0901586106.

Driscoll, Carlos A., Marilyn Menotti-Raymond, Alfred L. Roca, K. Hupe, W. E. Johnson, E. Geffen, E. H. Harley, et al. "The Near Eastern Origin of Cat Domestication." *Science* 317, no. 5837 (July 27, 2007): 519–523. https://doi.org/10.1126/science.1139518.

Dugatkin, Lee Alan, and Lyudmila Trut. *How to Tame a Fox (And Build a Dog).* (2017) Chicago, University of Chicago Press.

Dunsworth, Holly. "The Mismeasure of Dog." *The Mermaid's Tale* (blog), January 27, 2014. http://ecodevoevo.blogspot.com/2014/01/the-mismeasure-of-dog.html.

Francis, Richard. *Domesticated: Evolution in a Man-Made World.* (2015) New York, W. W. Norton.

Fugazza, Claudia, Ákos Pogány, and Ádám Miklósi. "Recall of Others' Actions after Incidental Encoding Reveals Episodic-like Memory in Dogs." *Current Biology* 26, no. 23 (December 2016): 3209–3213. https://doi.org/10.1016/j.cub.2016.09.057.

Gábor, A., M. Gácsi, D. Szabó, A. Miklósi, E. Kubinyi, and A. Andics. "Multilevel fMRI Adaptation for Spoken Word Processing in the Awake Dog Brain." *Scientific Reports* 10 (2020): 11968.

Gorman, James. "Dog Breeding in the Neolithic Age." *New York Times*, June 25, 2020. https://www.nytimes.com/2020/06/25/science/arctic-sled-dogs-genetics.html.

———. "The Leftovers Route to Dog Domestication." *New York Times*, January 7, 2021. https://www.nytimes.com/2021/01/07/science/wolves-dogs-domestication.html.

———. "What Wolf Pups That Play Fetch Reveal about Your Dog." *New York Times*, January 16, 2020. https://www.nytimes.com/2020/01/16/science/wolves-fetch-dogs-domestication.html.

———. "Why Are These Foxes Tame? Maybe They Weren't So Wild to Begin With." *New York Times*, December 3, 2019. https://www.nytimes.com/2019/12/03/science/foxes-tame-belyaev.html.

———. "Wolf Puppies Are Adorable. Then Comes the Call of the Wild." *New York Times*, October 13, 2017. https://www.nytimes.com/2017/10/13/science/wolves-dogs-genetics.html.

Grimm, David. "Dawn of the Dog." *Science* 348, no. 6232 (April 17, 2015): 274–279.

Handwerk, Brian. "How Accurate Is *Alpha*'s Theory of Dog Domestication?" *Smithsonian Magazine*, August 15, 2018. https://www.smithsonianmag.com/science-nature/how-wolves-really-became-dogs-180970014/.

Hansen Wheat, Christina, John L. Fitzpatrick, Björn Rogell, and Hans Temrin. "Behavioural Correlations of the Domestication Syndrome Are Decoupled in Modern Dog Breeds." *Nature Communications* 10, no. 1 (2019): 2422. https://doi.org/10.1038/s41467-019-10426-3.

Hansen Wheat, Christina, and Hans Temrin. "Intrinsic Ball Retrieving in Wolf Puppies Suggests Standing Ancestral Variation for Human-Directed Play Behavior." *iScience* 23, no. 2 (February 2020): 100811. https://doi.org/10.1016/j.isci.2019.100811.

Hekman, Jessica. "Research Updates." Darwin's Ark, January 10, 2019. https://darwinsark.org/forums/topic/research-updates/#post-7225.

Holson, Laura M. "Your Dog May Be Smart, but She's Not Exceptional." *New York Times*, October 8, 2018. https://www.nytimes.com/2018/10/08/science/dog-smart-study.html.

Johnson, Alexa. "Fact or Fiction: Foxes Were Brought to Catalina Island by the Native People." *The Island Naturalist*, no. 30/All about the Tongva. https://www.catalinaconservancy.org/index.php?s=news&p=article_333.

Kaminski, Juliane, Bridget M. Waller, Rui Diogo, Adam Hartstone-Rose, and Anne M. Burrows. "Evolution of Facial Muscle Anatomy in Dogs." *Proceedings of the National Academy of Sciences* 116, no. 29 (July 16, 2019): 14677–14681. https://doi.org/10.1073/pnas.1820653116.

Lahtinen, Maria, David Clinnick, Kristiina Mannermaa, J. Sakari Salonen, and Suvi Viranta. "Excess Protein Enabled Dog Domestication during Severe Ice Age Winters." *Scientific Reports* 11 (January 2021): 7. https://doi.org/10.1038/s41598-020-78214-4.

Lallensack, Rachael. "Ancient Genomes Heat Up Dog Domestication Debate." *Nature*, July 18, 2017. https://doi.org/10.1038/nature.2017.22320.

Lea, Stephen E. G., and Britta Osthaus. "In What Sense Are Dogs Special? Canine Cognition in Comparative Context." *Learning & Behavior* 46, no. 4 (December 2018): 335–363. https://doi.org/10.3758/s13420-018-0349-7.

Lord, Kathryn A., Greger Larson, Raymond P. Coppinger, and Elinor K. Karlsson. "The History of Farm Foxes Undermines the Animal Domestication Syndrome." *Trends in Ecology & Evolution* 35, no. 2 (February 2020): 125–136. https://doi.org/10.1016/j.tree.2019.10.011.

MacLean, Evan L., and Brian Hare. "Dogs Hijack the Human Bonding Pathway." *Science* 348, no. 6232 (April 17, 2015): 280–281. https://doi.org/10.1126/science.aab1200.

Magyari, L., Zs. Huszár, A. Turzó, and A. Andics. "Event-Related Potentials Reveal Limited Readiness to Access Phonetic Details during Word Processing in Dogs." *Royal Society Open Science* 7, no. 12 (December 2020): 200851. https://doi.org/10.1098/rsos.200851.

McConnell, Patricia B. "Canis Cousins: Unraveling Ancestral Ties." *The Bark Magazine*, Spring 2004.

———. "Gender Gap: Do Male and Female Dogs Learn Differently?" *The Bark Magazine*, June 2009.

———. "Point Taken: Have Dogs Evolved to Follow Our Lead?" *The Bark Magazine*, March 2011.

———. "Training Outside the Box: Can You Bet against Your Dog's Nature and Win?" *The Bark Magazine*, December 2008.

McDermott, Marie Tae. "The Vegan Dog." *New York Times*, June 6, 2017. https://www.nytimes.com/2017/06/06/well/family/the-vegan-dog.html.

McGreevy, Paul D., Dana Georgevsky, Johanna Carrasco, Michael Valenzuela, Deborah L. Duffy, and James A. Serpell. "Dog Behavior Co-Varies with Height, Bodyweight and Skull Shape." *PLoS ONE* 8, no. 12 (December 2013): e80529. https://doi.org/10.1371/journal.pone.0080529.

Morell, Virginia. "Dogs Understand Praise the Same Way We Do. Here's Why That Matters." *National Geographic*, August 6, 2020. https://www.nationalgeographic.com/animals/article/dogs-praise-brains-human-language.

Nagasawa, Miho, Shouhei Mitsui, Shiori En, Nobuyo Ohtani, Mitsuaki Ohta, Yasuo Sakuma, Tatsushi Onaka, Kazutaka Mogi, and Takefumi Kikusui. "Oxytocin-Gaze Positive Loop and the Coevolution of Human-Dog Bonds." *Science* 348, no. 6232 (April 17, 2015): 333–336.

"New Research Suggests Dogs Aren't Exceptionally Smart." *The Current*. CBC, October 15, 2018. https://www.cbc.ca/radio/thecurrent/the-current-for-october-15-2018-1.4862884/new-research-suggests-dogs-aren-t-exceptionally-smart-1.24862907.

Pennisi, Elizabeth. "Old Dogs Teach a New Lesson about Canine Origins." *Science*, November 15, 2013.

Pierotti, Raymond, and Brandy R. Fogg. *The First Domestication: How Wolves and Humans Coevolved.* (2017) New Haven, Yale University Press.

Range, Friederike, and Zsófia Virányi. "Social Learning from Humans or Conspecifics: Differences and Similarities between Wolves and Dogs." *Frontiers in Psychology* 4 (December 2013): 868. https://doi.org/10.3389/fpsyg.2013.00868.

Ratcliffe, Victoria F., and David Reby. "Orienting Asymmetries in Dogs' Responses to Different Communicatory Components of Human Speech." *Current Biology* 24, no. 24 (2014): 2908–2912. https://doi.org/10.1016/j.cub.2014.10.030.

Selba, Molly C., Emily R. Bryson, Ciele L. Rosenberg, Hock Gan Heng, and Valerie B. DeLeon. "Selective Breeding in Domestic Dogs: How Selecting for a Short Face Impacted Canine Neuroanatomy." *Anatomical Record* 304 (2020): 1–15. https://doi.org/10.1002/ar.24471.

Sinding, Mikkel-Holger S., Shyam Gopalakrishnan, Jazmín Ramos-Madrigal, Marc de Manuel, Vladimir V. Pitulko, Lukas Kuderna, Tatiana R. Feuerborn, et al. "Arctic-Adapted Dogs Emerged at the Pleistocene–Holocene Transition." *Science* 368, no. 6498 (June 26, 2020): 1495–1499. https://doi.org/10.1126/science.aaz8599.

Smith, Brett. "Dogs May Have Helped Early Man Hunt Mammoths to Extinction." *RedOrbit*, May 30, 2014.

Thalmann, O., B. Shapiro, P. Cui, V. J. Schuenemann, S. K. Sawyer, D. L. Greenfield, M. B. Germonpre, et al. "Complete Mitochondrial Genomes of Ancient Canids Suggest a European Origin of Domestic Dogs." *Science* 342, no. 6160 (November 15, 2013): 871–874. https://doi.org/10.1126/science.1243650.

Van Bourg, Joshua, Jordan Elizabeth Patterson, and Clive D. L. Wynne. "Pet Dogs (*Canis lupus familiaris*) Release Their Trapped and Distressed Owners: Individual Variation and Evidence of Emotional Contagion." *PLoS ONE* 15, no. 4 (2020): e0231742. https://doi.org/10.1371/journal.pone.0231742.

Viegas, Jennifer. "Dogs Play Dumb for Our Sake." *Discovery*, August 22, 2014. http://news.discovery.com/animals/pets/dogs-are-so-submissive-that-they-play-dumb-for-our-sake-140822.htm.

Wilkins, A. S., R. W. Wrangham, and W. T. Fitch. "The 'Domestication Syndrome' in Mammals: A Unified Explanation Based on Neural Crest Cell Behavior and Genetics." *Genetics* 197 (2014): 795–808.

Wrangham, Richard. *The Goodness Paradox: The Strange Relationship Between Virtue and Violence in Human Evolution*. (2019) New York, Pantheon.

Yong, Ed. "Can Dogs Smell Their 'Reflections'?" *The Atlantic*, August 17, 2017. https://www.theatlantic.com/science/archive/2017/08/can-dogs-smell-their -reflections/537219/.

———. "How Domestication Ruined Dogs' Pack Instincts." *The Atlantic*, October 16, 2017. https://www.theatlantic.com/science/archive/2017/10/how-domestication -ruined-dogs-pack-instincts/542994/.

Zak, Paul. "Dogs (and Cats) Can Love." *The Atlantic*, April 22, 2014. https://www .theatlantic.com/health/archive/2014/04/does-your-dog-or-cat-actually-love -you/360784/.

Zimmer, Carl. "A Virtual Pack, to Study Canine Minds." *New York Times*, April 22, 2013. https://www.nytimes.com/2013/04/23/science/enlisting-a-virtual-pack-to-study -dog-minds.html.

———. "Wolf to Dog: Scientists Agree on How, but Not Where." *New York Times*, November 14, 2013. https://www.nytimes.com/2013/11/14/science/wolf-to-dog -scientists-agree-on-how-but-not-where.html.

5. Wild-Mannered

Agnvall, B., A. Ali, S. Olby, and P. Jensen. "Red Junglefowl (*Gallus gallus*) Selected for Low Fear of Humans Are Larger, More Dominant and Produce Larger Offspring." *Animal* 8, no. 9 (2014): 1498–1505. https://doi.org/10.1017/S1751731114001426.

Barton, Loukas, Brittany Bingham, Krithivasan Sankaranarayanan, Cara Monroe, Ariane Thomas, and Brian M. Kemp. "The Earliest Farmers of Northwest China Exploited Grain-Fed Pheasants Not Chickens." *Scientific Reports* 10, no. 1 (February 2020): 2556. https://doi.org/10.1038/s41598-020-59316-5.

Buzz Beekeeping Supplies. "Bees Are Not Domestic Animals," n.d. http:// buzzbeekeepingsupplies.com/bees-are-not-domestic-animals/.

Berteselli, Greta Veronica, Barbara Regaiolli, Simona Normando, Barbara De Mori, Cesare Avesani Zaborra, and Caterina Spiezio. "European Wildcat and Domestic Cat: Do They Really Differ?" *Journal of Veterinary Behavior* 22 (November–December 2017): 35–40. https://doi.org/10.1016/j.jveb.2017.09.006.

Brust, Vera, and Anja Guenther. "Domestication Effects on Behavioural Traits and Learning Performance: Comparing Wild Cavies to Guinea Pigs." *Animal Cognition* 18, no. 1 (2015): 99–109. https://doi.org/10.1007/s10071-014-0781-9.

Callaway, Ewen. "Fowl Domination." *Nature* 515, no. 7528 (December 2014): 490–491. https://doi.org/10.1038/515490a.

———. "When Chickens Go Wild." *Nature* 529, no. 7586 (January 21, 2016): 270–273. https://doi.org/10.1038/529270a.

"Cats Are Famously Grumpy, but Do They Have Facial Expressions?" *New Scientist*, May 13, 2020. https://www.newscientist.com/lastword/mg24632821-300-cats-are -famously-grumpy-but-do-they-have-facial-expressions/.

Chen, Siyu, Chao Yan, Hai Xiang, Jinlong Xiao, Jian Liu, Hui Zhang, Jikun Wang, et al. "Transcriptome Changes Underlie Alterations in Behavioral Traits in Different Types of Chicken." *Journal of Animal Science* 98, no. 6 (2020): 1–10. https://doi .org/10.1093/jas/skaa167.

Coyne, Jerry. "New Date on First Domesticated Cats: Ca. 5300 Years Ago—and in China." *Why Evolution Is True* (blog), December 8, 2013. https://whyevolutionistrue .com/2013/12/18/new-date-on-first-domesticated-cats-ca-5300-years-ago-and-in -china/.

Dudde, Anissa, E. Tobias Krause, Lindsay R. Matthews, and Lars Schrader. "More Than Eggs—Relationship Between Productivity and Learning in Laying Hens." *Frontiers in Psychology* 9 (October 2018): 2000. https://doi.org/10.3389/fpsyg.2018.02000.

Engber, Daniel. "Test-Tube Piggies." *Slate*, June 18, 2012. https://slate.com/ technology/2012/06/human-guinea-pigs-and-the-history-of-the-iconic-lab-animal .html.

Frantz, Laurent A. F., Daniel G. Bradley, Greger Larson, and Ludovic Orlando. "Animal Domestication in the Era of Ancient Genomics." *Nature Reviews Genetics* 21, no. 8 (August 2020): 449–460. https://doi.org/10.1038/s41576-020-0225-0.

Gering, Eben, Darren Incorvaia, Rie Henriksen, Jeffrey Conner, Thomas Getty, and Dominic Wright. "Getting Back to Nature: Feralization in Animals and Plants." *Trends in Ecology & Evolution* 34, no. 12 (December 2019): 1137–1151. https://doi .org/10.1016/j.tree.2019.07.018.

Grandgeorge, Marine, Yentl Gautier, Yannig Bourreau, Heloise Mossu, and Martine Hausberger. "Visual Attention Patterns Differ in Dog vs. Cat Interactions with Children with Typical Development or Autism Spectrum Disorders." *Frontiers in Psychology* 11 (September 2020): 2047. https://doi.org/10.3389/fpsyg.2020.02047.

Grimm, David. "The Genes That Turned Wildcats into Kitty Cats." *Science* 346, no. 6211 (November 14, 2014): 799. https://doi.org/10.1126/science.346.6211.799.

Gut, Winnie, Lisa Crump, Jakob Zinsstag, Jan Hattendorf, and Karin Hediger. "The Effect of Human Interaction on Guinea Pig Behavior in Animal-Assisted Therapy." *Journal of Veterinary Behavior* 25 (May–June 2018): 56–64. https://doi.org/10.1016/j .jveb.2018.02.004.

Hansen Wheat, Christina, Wouter van der Bijl, and Christopher W. Wheat. "Morphology Does Not Covary with Predicted Behavioral Correlations of the Domestication Syndrome in Dogs." *Evolution Letters* 4, no. 3 (2020): 189–199. https://doi.org/10 .1002/evl3.168.

Henriksen, R., M. Johnsson, L. Andersson, P. Jensen, and D. Wright. "The Domesticated Brain: Genetics of Brain Mass and Brain Structure in an Avian Species." *Scientific Reports* 6, no. 1 (September 2016): 34031. https://doi.org/10.1038/srep34031.

Hu, Yaowu, Songmei Hu, Weilin Wang, Xiohong Wu, Fiona B. Marshall, Xianglong Chen, Liangliang Hou, and Changsui Wang. "Earliest Evidence for Commensal Processes of Cat Domestication." *Proceedings of the National Academy of Sciences* 111, no. 1 (January 2014): 116–120. https://doi.org/10.1073/pnas.1311439110.

Johnsson, M., E. Gering, P. Willis, S. Lopez, L. Van Dorp, G. Hellenthal, R. Henriksen, U. Friberg, and D. Wright. "Feralisation Targets Different Genomic Loci to Domestication in the Chicken." *Nature Communications* 7 (September 2016): 12950. https:// doi.org/10.1038/ncomms12950.

Kaiser, Sylvia, Michael B. Hennessy, and Norbert Sachser. "Domestication Affects the

Structure, Development and Stability of Biobehavioural Profiles." *Frontiers in Zoology* 12, no. Supplement 1 (2015): S19. https://doi.org/10.1186/1742-9994-12-S1-S19.

Kogan, Lori, and Shelly Volsche. "Not the Cat's Meow? The Impact of Posing with Cats on Female Perceptions of Male Dateability." *Animals* 10, no. 6 (June 9, 2020): 1007. https://doi.org/10.3390/ani10061007.

Krajcarz, Magdalena, Maciej T. Krajcarz, Mateusz Baca, Chris Baumann, Wim Van Neer, Danijela Popović, Magdalena Sudoł-Procyk, et al. "Ancestors of Domestic Cats in Neolithic Central Europe: Isotopic Evidence of a Synanthropic Diet." *Proceedings of the National Academy of Sciences* 117, no. 30 (2020): 17710–17719. https://doi.org/10.1073/pnas.1918884117.

Larson, Greger, and Dorian Q. Fuller. "The Evolution of Animal Domestication." *Annual Review of Ecology, Evolution, and Systematics* 45, no. 1 (2014): 115–136. https://doi.org/10.1146/annurev-ecolsys-110512-135813.

Lawler, Andrew. *Why Did the Chicken Cross the World? The Epic Saga of the Bird that Powers Civilization.* (2014) New York, Atria Books.

Lord, E., C. Collins, S. deFrance, M. J. LeFebvre, F. Pigière, P. Eeckhout, C. Erauw, et al. "Ancient DNA of Guinea Pigs (*Cavia* spp.) Indicates a Probable New Center of Domestication and Pathways of Global Distribution." *Scientific Reports* 10 (June 2020): 8901. https://doi.org/10.1038/s41598-020-65784-6.

Marshall, Fiona B., Keith Dobney, Tim Denham, and José M. Capriles. "Evaluating the Roles of Directed Breeding and Gene Flow in Animal Domestication." *Proceedings of the National Academy of Sciences* 111, no. 17 (April 2014): 6153–6158. https://doi.org/10.1073/pnas.1312984110.

Mehlhorn, Julia, and Stefanie Petow. "Smaller Brains in Laying Hens: New Insights into the Influence of Pure Breeding and Housing Conditions on Brain Size and Brain Composition." *Poultry Science* 99, no. 7 (2020): 3319–3327. https://doi.org/10.1016/j.psj.2020.03.039.

Montague, Michael J., Gang Li, Barbara Gandolfi, Razib Khan, Bronwen L. Aken, Steven M. J. Searle, Patrick Minx, et al. "Comparative Analysis of the Domestic Cat Genome Reveals Genetic Signatures Underlying Feline Biology and Domestication." *Proceedings of the National Academy of Sciences* 111, no. 48 (December 2014): 17230–17235. https://doi.org/10.1073/pnas.1410083111.

Newitz, Annalee. "Cats Are an Extreme Outlier among Domestic Animals." *Ars Technica*, June 19, 2017. https://arstechnica.com/science/2017/06/cats-are-an-extreme-outlier-among-domestic-animals/.

Offord, Scott. "Are Honey Bees Domesticated Livestock?" Keeping Backyard Bees, July 21, 2017. https://www.keepingbackyardbees.com/honey-bees-domesticated-livestock/.

Oldroyd, Benjamin P. "Corrigendum." *Molecular Ecology* 22, no. 5 (2013): 1483. https://doi.org/10.1111/mec.12133.

———. "Domestication of Honey Bees Was Associated with Expansion of Genetic Diversity." *Molecular Ecology* 21, no. 18 (2012): 4409–4411. https://doi.org/10.1111/j.1365-294X.2012.05641.x.

Ottoni, Claudio, Wim Van Neer, Bea De Cupere, Julien Daligault, Silvia Guimaraes, Joris Peters, Nikolai Spassov, et al. "The Palaeogenetics of Cat Dispersal in the

Ancient World." *Nature Ecology & Evolution* 1, no. 7 (2017): 0139. https://doi.org/10 .1038/s41559-017-0139.

Pennisi, Elizabeth. "Here's What Happens When You Rewind the Clock on Chicken Domestication." *Science*, September 15, 2015. https://doi.org/10.1126/science .aad1754.

Price, Edward O. *Animal Domestication and Behavior*. (2002) Wallingford, UK, CABI Publishing.

Ratliff, Evan. "Taming the Wild." *National Geographic*, March 2011. http://ngm .nationalgeographic.com/print/2011/03/taming-wild-animals/ratliff-text.

Ray, C. Claiborne. "The Wild Past of Domestic Cats." *New York Times*, August 19, 2013. https://www.nytimes.com/2013/08/20/science/the-wild-past-of-domestic -cats.html.

Sachser, Norbert, Michael B. Hennessy, and Sylvia Kaiser. "The Adaptive Shaping of Social Behavioural Phenotypes during Adolescence." *Biology Letters* 14, no. 11 (November 2018): 20180536. https://doi.org/10.1098/rsbl.2018.0536.

Seeley, Thomas D. "Who Were the Geniuses Who First Domesticated the Wild Honey Bee?" *Literary Hub*, May 24, 2019. https://lithub.com/who-were-the-geniuses-who -first-domesticated-the-wild-honey-bee/.

Silvestro, Daniele, Alexandre Antonelli, Nicolas Salamin, and Tiago B. Quental. "The Role of Clade Competition in the Diversification of North American Canids." *Proceedings of the National Academy of Sciences* 112, no. 28 (July 2015): 8684–8689. https://doi.org/10.1073/pnas.1502803112.

Taylor, Ashley P. "Domestication's Downsides for Dogs." *The Scientist*, December 21, 2015. http://www.the-scientist.com//?articles.view/articleNo/44889/title/ Domestication-s-Downsides-for-Dogs/.

Teletchea, Fabrice. "Animal Domestication: A Brief Overview." In *Animal Domestication*, edited by Fabrice Teletchea, 1–19. (2019) London, IntechOpen. https://doi.org/10 .5772/intechopen.86783.

University of Otago. "Origins of the Beloved Guinea Pig." *ScienceDaily*, June 16, 2020. https://www.sciencedaily.com/releases/2020/06/200616100818.htm.

Wang, Ming-Shan, Mukesh Thakur, Min-Sheng Peng, Yu Jiang, Laurent Alain François Frantz, Ming Li, Jin-Jin Zhang, et al. "863 Genomes Reveal the Origin and Domestication of Chicken." *Cell Research* 30, no. 8 (2020): 693–701. https://doi.org/10.1038/ s41422-020-0349-y.

Zimmermann, Tobias D., Sylvia Kaiser, Michael B. Hennessy, and Norbert Sachser. "Adaptive Shaping of the Behavioural and Neuroendocrine Phenotype during Adolescence." *Proceedings of the Royal Society B: Biological Sciences* 284, no. 1849 (2017): 20162784. https://doi.org/10.1098/rspb.2016.2784.

6. The Anxious Invertebrate

Adriaens, Pieter R., and Andreas De Block. "The Evolutionary Turn in Psychiatry: A Historical Overview." *History of Psychiatry* 21, no. 2 (2010): 131–143. https://doi .org/10.1177/0957154X10370632.

Allen, Nicholas B., and Paul B. T. Badcock. "Darwinian Models of Depression: A Review of Evolutionary Accounts of Mood and Mood Disorders." *Progress in Neuro-Psycho pharmacology and Biological Psychiatry* 30, no. 5 (2006): 815–826. https://doi.org/10 .1016/j.pnpbp.2006.01.007.

Bacqué-Cazenave, Julien, Marion Berthomieu, Daniel Cattaert, Pascal Fossat, and Jean Paul Delbecque. "Do Arthropods Feel Anxious during Molts?" *Journal of Experimental Biology* 222, no. 2 (January 1, 2018): jeb.186999. https://doi.org/10.1242/jeb .186999.

Bacqué-Cazenave, Julien, Rahul Bharatiya, Grégory Barrière, Jean-Paul Delbecque, Nouhaila Bouguiyoud, Giuseppe Di Giovanni, Daniel Cattaert, and Philippe De Deurwaerdère. "Serotonin in Animal Cognition and Behavior." *International Journal of Molecular Sciences* 21, no. 5 (February 28, 2020): 1649. https://doi.org/10.3390/ ijms21051649.

Bacqué-Cazenave, Julien, Daniel Cattaert, Jean-Paul Delbecque, and Pascal Fossat. "Social Harassment Induces Anxiety-like Behaviour in Crayfish." *Scientific Reports* 7 (February 2017): 39935. https://doi.org/10.1038/srep39935.

Beyer, Dominik K. E., and Nadja Freund. "Animal Models for Bipolar Disorder: From Bedside to the Cage." *International Journal of Bipolar Disorders* 5 (December 2017): 35. https://doi.org/10.1186/s40345-017-0104-6.

Bielecka, Krystyna, and Mira Marcinów. "Mental Misrepresentation in Non-Human Psychopathology." *Biosemiotics* 10, no. 2 (2017): 195–210. https://doi.org/10.1007/ s12304-017-9299-2.

Boer, Sietse F. de, Bauke Buwalda, and Jaap M. Koolhaas. "Untangling the Neurobiology of Coping Styles in Rodents: Towards Neural Mechanisms Underlying Individual Differences in Disease Susceptibility." *Neuroscience and Biobehavioral Reviews* 74, Part B (March 2017): 401–422. https://doi.org/10.1016/j.neubiorev.2016.07.008.

Border, Richard, Emma C. Johnson, Luke M. Evans, Andrew Smolen, Noah Berley, Patrick F. Sullivan, and Matthew C. Keller. "No Support for Historical Candidate Gene or Candidate Gene-by-Interaction Hypotheses for Major Depression across Multiple Large Samples." *American Journal of Psychiatry* 176, no. 5 (May 2019): 376–387. https://doi.org/10.1176/appi.ajp.2018.18070881.

Braitman, Laurel. *Animal Madness: Inside Their Minds*. (2014) New York, Simon and Schuster Paperbacks.

Bryn, Brandon. "Crayfish Can Be Calmed with Anti-Anxiety Medication." AAAS.org, June 11, 2014. https://www.aaas.org/news/science-crayfish-can-be-calmed-anti -anxiety-medication.

Choi, Charles. "Could Human Evolutionary Changes Be behind Mental Disorders?" *Discover*, August 9, 2018. https://www.discovermagazine.com/mind/could-human -evolutionary-changes-be-behind-mental-disorders.

Cook, Gareth. "Psychiatry for Animals." *Scientific American*, September 6, 2016. https:// www.scientificamerican.com/article/psychiatry-for-animals/.

Dasgupta, Shreya. "Many Animals Can Become Mentally Ill." BBC, September 9, 2015. http://www.bbc.com/earth/story/20150909-many-animals-can-become-mentally -ill.

Dodman, Nicholas. *Pets on the Couch: Neurotic Dogs, Compulsive Cats, Anxious Birds, and the New Science of Animal Psychiatry*. (2016) New York, Atria Books.

Durisko, Zachary, Benoit H. Mulsant, Kwame McKenzie, and Paul W. Andrews. "Using Evolutionary Theory to Guide Mental Health Research." *Canadian Journal of Psychiatry* 61, no. 3 (2016): 159–165. https://doi.org/10.1177/0706743716632517.

Eisenberg, Dan T. A., Benjamin Campbell, Peter B. Gray, and Michael D. Sorenson. "Dopamine Receptor Genetic Polymorphisms and Body Composition in Undernourished Pastoralists: An Exploration of Nutrition Indices among Nomadic and Recently Settled Ariaal Men of Northern Kenya." *BMC Evolutionary Biology* 8 (June 2008): 173. https://doi.org/10.1186/1471-2148-8-173.

Fossat, Pascal, Julien Bacque-Cazenave, Philippe De Deurwaerdere, Jean-Paul Delbecque, and Daniel Cattaert. "Anxiety-like Behavior in Crayfish Is Controlled by Serotonin." *Science* 344, no. 6189 (June 13, 2014): 1293–1297. https://doi.org/10.1126/science.1248811.

Frances, Allen J. "What You Need to Know about the Genetics of Mental Disorders." *Psychology Today*, April 30, 2016. https://www.psychologytoday.com/us/blog/saving-normal/201604/what-you-need-know-about-the-genetics-mental-disorders.

French, Jeffrey A. "The Marmoset as a Model in Behavioral Neuroscience and Psychiatric Research." In *The Common Marmoset in Captivity and Biomedical Research*, edited by Robert P. Marini, Lynn M. Wachtman, Suzette D. Tardif, Keith Mansfield, and James G. Fox, 477–491. (2019) London, Academic Press. https://doi.org/10.1016/B978-0-12-811829-0.00026-1.

Gordon, Joshua. "What Can Animals Tell Us about Mental Illnesses?" National Institute of Mental Health, October 21, 2019. https://www.nimh.nih.gov/about/director/messages/2019/what-can-animals-tell-us-about-mental-illnesses.

Gururajan, Anand, Andreas Reif, John F. Cryan, and David A. Slattery. "The Future of Rodent Models in Depression Research." *Nature Reviews Neuroscience* 20, no. 11 (November 2019): 686–701. https://doi.org/10.1038/s41583-019-0221-6.

Hassan, Ahmad. "Animal Psychopathology." In *Encyclopedia of Animal Cognition and Behavior*, edited by Jennifer Vonk and Todd Shackelford, 1–8. (2020) Cham, Switzerland: Springer. https://doi.org/10.1007/978-3-319-47829-6_2093-1.

Janiak, M. C., S. L. Pinto, G. Duytschaever, M. A. Carrigan, and A. D. Melin. "Genetic Evidence of Widespread Variation in Ethanol Metabolism among Mammals: Revisiting the 'Myth' of Natural Intoxication." *Biology Letters* 16 (2020): 20200070.

Kaiser, Tobias, and Guoping Feng. "Modeling Psychiatric Disorders for Developing Effective Treatments." *Nature Medicine* 21, no. 9 (September 2015): 979–988. https://doi.org/10.1038/nm.3935.

Korneliussen, Ida. "Can Wild Animals Have Mental Illnesses?" *ScienceNorway*, June 24, 2015. https://sciencenorway.no/animal-behaviour-animal-kingdom-domestication/can-wild-animals-have-mental-illnesses/1417928.

Moffitt, Matthew, and Himanshu Sharma. "5 Brain Disorders That Started as Evolutionary Advantages." Cracked, March 6, 2014. https://www.cracked.com/article_20905_5-brain-disorders-that-started-as-evolutionary-advantages.html.

Nesse, Randolph M. *Good Reasons for Bad Feelings: Insights from the Frontier of Evolutionary Psychiatry*. (2019) New York, Penguin Random House.

Noh, Hyun Ji, Ruqi Tang, Jason Flannick, Colm O'Dushlaine, Ross Swofford, Daniel Howrigan, Diane P. Genereux, et al. "Integrating Evolutionary and Regulatory

Information with a Multispecies Approach Implicates Genes and Pathways in Obsessive-Compulsive Disorder." *Nature Communications* 8, no. 1 (December 2017): 774. https://doi.org/10.1038/s41467-017-00831-x.

Price, John. "The Dominance Hierarchy and the Evolution of Mental Illness." *The Lancet* 290, no. 7509 (July 1967): 243–246. https://doi.org/10.1016/S0140-6736(67)92306-9.

Reardon, Sara. "How Evolution Has Shaped Mental Illness." *Nature*, November 2, 2017.

Scott, Daniel, and Carol A. Tamminga. "Effects of Genetic and Environmental Risk for Schizophrenia on Hippocampal Activity and Psychosis-like Behavior in Mice." *Behavioural Brain Research* 339 (February 2018): 114–123. https://doi.org/10.1016/j.bbr.2017.10.039.

Song, Janet H. T., Craig B. Lowe, and David M. Kingsley. "Characterization of a Human-Specific Tandem Repeat Associated with Bipolar Disorder and Schizophrenia." *American Journal of Human Genetics* 103, no. 3 (September 2018): 421–430. https://doi.org/10.1016/j.ajhg.2018.07.011.

Stetka, Bret. "Why Don't Animals Get Schizophrenia (and How Come We Do)?" *Scientific American*, March 24, 2015. https://www.scientificamerican.com/article/why-don-t-animals-get-schizophrenia-and-how-come-we-do/.

Wilkins, Adam S., Richard W. Wrangham, and W. Tecumseh Fitch. "The 'Domestication Syndrome' in Mammals: A Unified Explanation Based on Neural Crest Cell Behavior and Genetics." *Genetics* 197, no. 3 (July 2014): 795–808. https://doi.org/10.1534/genetics.114.165423.

Wolmarans, De Wet, Isabella M. Scheepers, Dan J. Stein, and Brian H. Harvey. "*Peromyscus maniculatus bairdii* as a Naturalistic Mammalian Model of Obsessive-Compulsive Disorder: Current Status and Future Challenges." *Metabolic Brain Disease* 33, no. 2 (April 2018): 443–455. https://doi.org/10.1007/s11011-017-0161-7.

Yong, Ed. "A Waste of 1,000 Research Papers." *The Atlantic*, May 17, 2019. https://www.theatlantic.com/science/archive/2019/05/waste-1000-studies/589684/.

7. Dancing Cockatoos and Thieving Gulls

Ashton, Benjamin J., Patrick Kennedy, and Andrew N. Radford. "Interactions with Conspecific Outsiders as Drivers of Cognitive Evolution." *Nature Communications* 11, no. 1 (October 2020): 4937. https://doi.org/10.1038/s41467-020-18780-3.

Auersperg, A. M. I., C. Köck, A. Pledermann, M. O'Hara, and L. Huber. "Safekeeping of Tools in Goffin's Cockatoos, *Cacatua goffiniana*." *Animal Behaviour* 128 (June 2017): 125–133. https://doi.org/10.1016/j.anbehav.2017.04.010.

Bastos, Amalia P. M., and Alex H. Taylor. "Kea Show Three Signatures of Domain-General Statistical Inference." *Nature Communications* 11, no. 1 (December 2020): 828. https://doi.org/10.1038/s41467-020-14695-1.

Bentley-Condit, Vicki K., and E. O. Smith. "Animal Tool Use: Current Definitions and an Updated Comprehensive Catalog." *Behaviour* 147, no. 2 (2010): 185–221. https://doi.org/10.1163/000579509X12512865686555.

Boeckle, Markus, and Nicola S. Clayton. "A Raven's Memories Are for the Future." *Science* 357, no. 6347 (July 14, 2017): 126–127. https://doi.org/10.1126/science.aan8802.

Bugnyar, Thomas. "Tool Use: New Caledonian Crows Engage in Mental Planning." *Cur-

rent Biology 29, no. 6 (March 2019): R200–202. https://doi.org/10.1016/j.cub.2019 .01.059.

Cairó, Osvaldo. "External Measures of Cognition." *Frontiers in Human Neuroscience* 5 (October 2011): 108. https://doi.org/10.3389/fnhum.2011.00108.

Cassella, Carly. "Gulls Work Out the Timing of School Lunch Breaks So They Can Steal Food." ScienceAlert, November 13, 2020. https://www.sciencealert.com/gulls-in-the -uk-have-figured-out-when-schools-are-on-lunch-break-so-they-can-steal-food.

Ducatez, Simon, Daniel Sol, Ferran Sayol, and Louis Lefebvre. "Behavioural Plasticity Is Associated with Reduced Extinction Risk in Birds." *Nature Ecology & Evolution* 4, no. 6 (June 2020): 788–793. https://doi.org/10.1038/s41559-020-1168-8.

Elbein, Asher. "These Birds Eat Fire, or Close to It, to Live Another Day." *New York Times*, April 28, 2020. https://www.nytimes.com/2020/04/28/science/birds-diets -survival.html.

"Encephalization Quotient." Psychology Wiki. Accessed November 29, 2020. https:// psychology.wikia.org/wiki/Encephalization_quotient.

Fine, M. L., M. H. Horn, and B. Cox. "*Acanthonus armatus*, a Deep-Sea Teleost Fish with a Minute Brain and Large Ears." *Proceedings of the Royal Society B: Biological Sciences* 230, no. 1259 (March 1987): 257–265. https://doi.org/10.1098/rspb.1987.0018.

Gorman, James. "Trash Parrots Invent New Skill in Australian Suburbs." *New York Times*, July 22, 2021. https://www.nytimes.com/2021/07/22/science/trash-parrots -australia.html.

———. "Wild Cockatoos Are Just as Smart as Lab-Raised Ones." *New York Times*, May 26, 2020. https://www.nytimes.com/2020/05/26/science/cockatoos-intelligence .html.

GrrlScientist. "Parrot Genomes Provide Insights into Evolution of Longevity and Cognition." *Dialogue & Discourse* (blog), December 20, 2018. https://medium.com/ discourse/parrot-genomes-provide-insights-into-evolution-of-longevity-and -cognition-c9ddf3396361.

Gruber, Romana, Martina Schiestl, Markus Boeckle, Anna Frohnwieser, Rachael Miller, Russell D. Gray, Nicola S. Clayton, and Alex H. Taylor. "New Caledonian Crows Use Mental Representations to Solve Metatool Problems." *Current Biology* 29, no. 4 (February 2019): 686–692. https://doi.org/10.1016/j.cub.2019.01.008.

Hampton, Robert. "Parallel Overinterpretation of Behavior of Apes and Corvids." *Learning & Behavior* 47, no. 2 (June 2019): 105–106. https://doi.org/10.3758/s13420 -018-0330-5.

Hattori, Yuko, and Masaki Tomonaga. "Rhythmic Swaying Induced by Sound in Chimpanzees (*Pan troglodytes*)." *Proceedings of the National Academy of Sciences* 117, no. 2 (January 14, 2020): 936–942. https://doi.org/10.1073/pnas.1910318116.

Herculano-Houzel, Suzana. "Birds Do Have a Brain Cortex—and Think." *Science* 369, no. 6511 (September 25, 2020): 1567–1568. https://doi.org/10.1126/science.abe0536.

Hooper, Rowan. "Chimps Have Local Culture Differences When It Comes to Eating Termites." *New Scientist*, May 28, 2020. https://www.newscientist.com/article/2244799 -chimps-have-local-culture-differences-when-it-comes-to-eating-termites/.

Hunt, Gavin R. "Manufacture and Use of Hook-Tools by New Caledonian Crows." *Nature* 379 (January 18, 1996): 249–251. https://doi.org/10.1038/379249a0.

Jao Keehn, R. Joanne, John R. Iversen, Irena Schulz, and Aniruddh D. Patel. "Spontaneity and Diversity of Movement to Music Are Not Uniquely Human." *Current Biology* 29, no. 13 (July 2019): R621–622. https://doi.org/10.1016/j.cub.2019 .05.035.

Jarvis, Erich D., Onur Güntürkün, Laura Bruce, András Csillag, Harvey Karten, Wayne Kuenzel, Loreta Medina, et al. "Avian Brains and a New Understanding of Vertebrate Brain Evolution." *Nature Reviews Neuroscience* 6, no. 2 (February 2005): 151–159. https://doi.org/10.1038/nrn1606.

Jeong, Sarah, and Rachel Becker. "Science Doesn't Explain Tech's Diversity Problem—History Does." *The Verge*, August 16, 2017. https://www.theverge .com/2017/8/16/16153740/tech-diversity-problem-science-history-explainer -inequality.

Jiménez-Ortega, Dante, Niclas Kolm, Simone Immler, Alexei A. Maklakov, and Alejandro Gonzalez-Voyer. "Long Life Evolves in Large-Brained Bird Lineages." *Evolution* 74, no. 12 (December 2020): 2617–2628. https://doi.org/10.1111/evo.14087.

Kabadayi, Can, and Mathias Osvath. "Ravens Parallel Great Apes in Flexible Planning for Tool-Use and Bartering." *Science* 357, no. 6347 (July 14, 2017): 202–204. https:// doi.org/10.1126/science.aam8138.

Kim, Shi En. "Why Australia's Trash Bin–Raiding Cockatoos Are the 'Punks of the Bird World.'" *Smithsonian Magazine*, July 22, 2021. https://www.smithsonianmag .com/science-nature/cockatoos-learn-open-garbage-bins-observing-their-peers -180978248/.

Klump, Barbara C., John M. Martin, Sonja Wild, Jana K. Hörsch, Richard E. Major, and Lucy M. Aplin. "Innovation and Geographic Spread of a Complex Foraging Culture in an Urban Parrot." *Science* 373, no. 6553 (July 23, 2021): 456–460. https://doi .org/10.1126/science.abe7808.

Kohda, M., T. Hottal, T. Takeyama, S. Awata, H.Tanaka, J-Y. Asai, and A. L. Jordan. "If a Fish Can Pass the Mark Test, What Are the Implications for Consciousness and Self-Awareness Testing in Animals?" *PLoS Biology* 17(2): e3000021.

Ksepka, Daniel T., Amy M. Balanoff, N. Adam Smith, Gabriel S. Bever, Bhart-Anjan S. Bhullar, Estelle Bourdon, Edward L. Braun, et al. "Tempo and Pattern of Avian Brain Size Evolution." *Current Biology* 30, no. 11 (June 2020): 2026–2036.e3. https://doi .org/10.1016/j.cub.2020.03.060.

Laumer, I. B., S. A. Jelbert, A. H. Taylor, T. Rössler, and A. M. I. Auersperg. "Object Manufacture Based on a Memorized Template: Goffin's Cockatoos Attend to Different Model Features." *Animal Cognition* 24, no. 11 (May 2021): 457–470. https://doi .org/10.1007/s10071-020-01435-7.

Lefebvre, Louis, Patrick Whittle, Evan Lascaris, and Adam Finkelstein. "Feeding Innovations and Forebrain Size in Birds." *Animal Behaviour* 53, no. 3 (March 1997): 549–560. https://doi.org/10.1006/anbe.1996.0330.

McCoy, Dakota E., Martina Schiestl, Patrick Neilands, Rebecca Hassall, Russell D. Gray, and Alex H. Taylor. "New Caledonian Crows Behave Optimistically after Using Tools." *Current Biology* 29, no. 16 (August 2019): 2737-2742.e3. https://doi.org/10 .1016/j.cub.2019.06.080.

Nicolakakis, Nektaria, and Louis Lefebvre. "Forebrain Size and Innovation Rate in

European Birds: Feeding, Nesting and Confounding Variables." *Behaviour* 137, no. 11 (2000): 1415–1429. https://doi.org/10.1163/156853900502646.

Patel, Aniruddh D., John R. Iversen, Micah R. Bregman, and Irena Schulz. "Experimental Evidence for Synchronization to a Musical Beat in a Nonhuman Animal." *Current Biology* 19, no. 10 (May 2009): 827–830. https://doi.org/10.1016/j.cub.2009.03.038.

Reader, Simon M., and Kevin N. Laland. "Social Intelligence, Innovation, and Enhanced Brain Size in Primates." *Proceedings of the National Academy of Sciences* 99, no. 7 (April 2, 2002): 4436–4441. https://doi.org/10.1073/pnas.062041299.

Reiner, Anton. "An Explanation of Behavior." Review of *The Triune Brain in Evolution: Role in Paleocerebral Functions*, by Paul D. MacLean. *Science*, n.s., 250, no. 4978 (October 12, 1990): 303–305. https://doi.org/10.1126/science.250.4978.303-a.

Rössler, Theresa, Berenika Mioduszewska, Mark O'Hara, Ludwig Huber, Dewi M. Prawiradilaga, and Alice M. I. Auersperg. "Using an Innovation Arena to Compare Wild-Caught and Laboratory Goffin's Cockatoos." *Scientific Reports* 10, no. 1 (December 2020): 8681. https://doi.org/10.1038/s41598-020-65223-6.

Rozsa, Matthew. "No, You Don't Have a 'Lizard Brain': Why the Psychology 101 Model of the Brain Is All Wrong." *Salon*, May 17, 2020. https://www.salon.com/2020/05/17/no-you-dont-have-a-lizard-brain-why-the-psychology-101-model-of-the-brain-is-all-wrong/.

Sayol, Ferran, Oriol Lapiedra, Simon Ducatez, and Daniel Sol. "Larger Brains Spur Species Diversification in Birds." *Evolution* 73, no. 10 (October 2019): 2085–2093. https://doi.org/10.1111/evo.13811.

Sayol, Ferran, Joan Maspons, Oriol Lapiedra, Andrew N. Iwaniuk, Tamás Székely, and Daniel Sol. "Environmental Variation and the Evolution of Large Brains in Birds." *Nature Communications* 7, no. 1 (December 2016): 13971. https://doi.org/10.1038/ncomms13971.

Schulze-Makuch, Dirk. "The Naked Mole-Rat: An Unusual Organism with an Unexpected Latent Potential for Increased Intelligence?" *Life* 9, no. 3 (2019): 76. https://doi.org/10.3390/life9030076.

Shultz, S. "Brain Evolution in Vertebrates." In *Encyclopedia of Behavioral Neuroscience*, edited by George Koob, Michelle Le Moal, and Richard Thompson, 1st ed., 180–186. (2010) London, Academic Press.

Snell-Rood, E. C., and N. Wick. "Anthropogenic Environments Exert Variable Selection on Cranial Capacity in Mammals." *Proceedings of the Royal Society B* (2013) 280: 20131384.

Sol, D. "Revisiting the Cognitive Buffer Hypothesis for the Evolution of Large Brains." *Biology Letters* 5 (2009): 130–133.

Sorato, Enrico, Josefina Zidar, Laura Garnham, Alastair Wilson, and Hanne Løvlie. "Heritabilities and Co-Variation among Cognitive Traits in Red Junglefowl." *Philosophical Transactions of the Royal Society B: Biological Sciences* 373, no. 1756 (September 2018): 20170285. https://doi.org/10.1098/rstb.2017.0285.

Spelt, Anouk, Oliver Soutar, Cara Williamson, Jane Memmott, Judy Shamoun-Baranes, Peter Rock, and Shane Windsor. "Urban Gulls Adapt Foraging Schedule to Human-Activity Patterns." *Ibis* 163, no. 1 (January 2021): 274–282. https://doi.org/10.1111/ibi.12892.

Stacho, Martin, Christina Herold, Noemi Rook, Hermann Wagner, Markus Axer, Katrin Amunts, and Onur Güntürkün. "A Cortex-like Canonical Circuit in the Avian Forebrain." *Science* 369, no. 6511 (September 25, 2020): eabc5534. https://doi.org/10.1126/science.abc5534.

Triki, Zegni, Yasmin Emery, Magda C. Teles, Rui F. Oliveira, and Redouan Bshary. "Brain Morphology Predicts Social Intelligence in Wild Cleaner Fish." *Nature Communications* 11, no. 1 (December 2020): 6423. https://doi.org/10.1038/s41467-020-20130-2.

Uomini, Natalie, Joanna Fairlie, Russell D. Gray, and Michael Griesser. "Extended Parenting and the Evolution of Cognition." *Philosophical Transactions of the Royal Society B: Biological Sciences* 375, no. 1803 (July 2020): 20190495. https://doi.org/10.1098/rstb.2019.0495.

Vonk, Jennifer. "Sticks and Stones: Associative Learning Alone?" *Learning & Behavior* 48, no. 3 (September 2020): 277–278. https://doi.org/10.3758/s13420-019-00387-4.

"Why Is There Only One Intelligent Species? Why Aren't Other Species Intelligent?" Quora. Accessed September 14, 2017. https://www.quora.com/Why-is-there-only-one-intelligent-species-Why-aren%E2%80%99t-other-species-intelligent?share=1.

Wirthlin, Morgan, Nicholas C. B. Lima, Rafael Lucas Muniz Guedes, André E. R. Soares, Luiz Gonzaga P. Almeida, Nathalia P. Cavaleiro, Guilherme Loss de Morais, et al. "Parrot Genomes and the Evolution of Heightened Longevity and Cognition." *Current Biology* 28, no. 24 (December 2018): 4001–4008.e7. https://doi.org/10.1016/j.cub.2018.10.050.

8. A Soft Spot for Hard Creatures

Aldhous, Peter. "Zoologger: My Brain's So Big It Spills into My Legs." *New Scientist*, December 14, 2011. https://www.newscientist.com/article/dn21285-zoologger-my-brains-so-big-it-spills-into-my-legs/.

Amodio, Piero, Markus Boeckle, Alexandra K. Schnell, Ljerka Ostojíc, Graziano Fiorito, and Nicola S. Clayton. "Grow Smart and Die Young: Why Did Cephalopods Evolve Intelligence?" *Trends in Ecology & Evolution* 34, no. 1 (January 2019): 45–56. https://doi.org/10.1016/j.tree.2018.10.010.

Bailey, Nathan W., and Marlene Zuk. "Field Crickets Change Mating Preferences Using Remembered Social Information." *Biology Letters* 5 (2009): 449–451.

Balter, Michael. "'Killjoys' Challenge Claims of Clever Animals." *Science* 335, no. 6072 (March 2012): 1036–1037. https://doi.org/10.1126/science.335.6072.1036.

Bandini, Elisa, Alba Motes-Rodrigo, Matthew P. Steele, Christian Rutz, and Claudio Tennie. "Examining the Mechanisms Underlying the Acquisition of Animal Tool Behaviour." *Biology Letters* 16, no. 6 (June 2020): 20200122. https://doi.org/10.1098/rsbl.2020.0122.

Bastos, Amalia P. M., and Alex H. Taylor. "Macphail's Null Hypothesis of Vertebrate Intelligence: Insights from Avian Cognition." *Frontiers in Psychology* 11 (July 2020): 1692. https://doi.org/10.3389/fpsyg.2020.01692.

Carrington, Damian. "Honey Bees Use Animal Poo to Repel Giant Hornet Attacks." *Guardian*, December 9, 2020. https://www.theguardian.com/environment/2020/dec/09/honey-bees-use-animal-poo-to-repel-giant-hornet-attacks.

Chandra, Vikram, Ingrid Fetter-Pruneda, Peter R. Oxley, Amelia L. Ritger, Sean K. McKenzie, Romain Libbrecht, and Daniel J. C. Kronauer. "Social Regulation of Insulin Signaling and the Evolution of Eusociality in Ants." *Science* 361, no. 6400 (July 2018): 398–402. https://doi.org/10.1126/science.aar5723.

Chittka, Lars. "The Secret Lives of Bees as Horticulturists?" *Science* 368, no. 6493 (May 2020): 824–825. https://doi.org/10.1126/science.abc2451.

Chittka, Lars, and Jeremy Niven. "Are Bigger Brains Better?" *Current Biology* 19, no. 21 (November 2009): R995–1008. https://doi.org/10.1016/j.cub.2009.08.023.

Coyne, Jerry. "Tool-Using Ants Build Siphons to Wick Sugar Water Out of Containers, Keeping Them from Drowning." *Why Evolution Is True* (blog), October 11, 2020. https://whyevolutionistrue.com/2020/10/11/tool-using-ants-build-siphons-to-wick-sugar-water-out-of-containers-so-they-dont-drown/.

Cross, Fiona R., Georgina E. Carvell, Robert R. Jackson, and Randolph C. Grace. "Arthropod Intelligence? The Case for *Portia*." *Frontiers in Psychology* 11 (October 2020): 568049. https://doi.org/10.3389/fpsyg.2020.568049.

Darwin, C. *The Descent of Man and Selection in Relation to Sex.* (1871) London, John Murray.

Davies, Ross, Mary H. Gagen, James C. Bull, and Edward C. Pope. "Maze Learning and Memory in a Decapod Crustacean." *Biology Letters* 15, no. 10 (October 2019): 20190407. https://doi.org/10.1098/rsbl.2019.0407.

Dennett, Daniel C. "Animal Consciousness: What Matters and Why." *Social Research* 62, no. 3 (Fall 1995): 691–710.

Dockrill, Peter. "Octopuses Observed Punching Fish, Perhaps Out of Spite, Scientists Say." ScienceAlert, December 21, 2020. https://www.sciencealert.com/octopuses-observed-punching-fish-perhaps-out-of-spite-scientists-say.

Dvorsky, George. "Bees Can Learn Symbols Associated with Counting, New Experiment Suggests." Gizmodo, June 5, 2019. https://gizmodo.com/bees-can-learn-symbols-associated-with-counting-new-ex-1835245766.

Eberhard, William G. "Are Smaller Animals Behaviourally Limited? Lack of Clear Constraints in Miniature Spiders." *Animal Behaviour* 81, no. 4 (April 2011): 813–823. https://doi.org/10.1016/j.anbehav.2011.01.016.

Eberhard, William G., and William T. Wcislo. "Plenty of Room at the Bottom?" *American Scientist* 100 (June 2012): 226–233.

Flaim, Mary, and Aaron P. Blaisdell. "The Comparative Analysis of Intelligence." *Psychological Bulletin* 146, no. 12 (December 2020): 1174–1199. https://doi.org/10.1037/bul0000306.

Frasnelli, Elisa, Théo Robert, Pizza Ka Yee Chow, Ben Scales, Sam Gibson, Nicola Manning, Andrew O. Philippides, Thomas S. Collett, and Natalie Hempel de Ibarra. "Small and Large Bumblebees Invest Differently When Learning about Flowers." *Current Biology* 31, no. 5 (March 2021): 1058–1064.e3. https://doi.org/10.1016/j.cub.2020.11.062.

Frishberg, Hannah. "Octopuses Spite-Punch Fish, Who 'Don't Like It,' Study Finds." *New York Post*, December 23, 2020. https://nypost.com/2020/12/23/octopuses-spite-punch-fish-who-dont-like-it-study/.

Godfrey-Smith, Peter. *Metazoa: Animal Life and the Birth of the Mind.* (2020) New York, Farrar, Straus and Giraux.

———. *Other Minds: The Octopus, the Sea, and the Deep Origins of Consciousness.* (2016) New York, Farrar, Straus and Giraux.

Greenwood, Veronique. "Cuttlefish Took Something like a Marshmallow Test. Many Passed." *New York Times*, December 30, 2020. https://www.nytimes.com/2020/12/30/science/cuttlefish-counting.html.

Greshko, Michael. "Brainless Creatures Can Do Some Incredibly Smart Things." *National Geographic*, May 21, 2018. https://www.nationalgeographic.com/science/article/life-without-brains-smart-slime-molds-plants-jellyfish-osr-science.

Grüter, Christoph, and Tomer J. Czaczkes. "Communication in Social Insects and How It Is Shaped by Individual Experience." *Animal Behaviour* 151 (May 2019): 207–215. https://doi.org/10.1016/j.anbehav.2019.01.027.

Hare, Robin M., and Leigh W. Simmons. "Sexual Selection and Its Evolutionary Consequences in Female Animals." *Biological Reviews* 94, no. 3 (June 2019): 929–956. https://doi.org/10.1111/brv.12484.

Hays, Brooks. "Large Bumblebees Memorize Locations of Biggest, Best Flowers." *UPI*, December 28, 2020. https://www.upi.com/Science_News/2020/12/28/Large-bumblebees-memorize-locations-of-biggest-best-flowers/7201609173223/.

Hernandez-Lallement, Julen, Marijn van Wingerden, and Tobias Kalenscher. "Towards an Animal Model of Callousness." *Neuroscience and Biobehavioral Reviews* 91 (August 2018): 121–129. https://doi.org/10.1016/j.neubiorev.2016.12.029.

Howard, Scarlett R., Aurore Avarguès-Weber, Jair E. Garcia, Andrew D. Greentree, and Adrian G. Dyer. "Symbolic Representation of Numerosity by Honeybees (*Apis mellifera*): Matching Characters to Small Quantities." *Proceedings of the Royal Society B: Biological Sciences* 286, no. 1904 (June 2019): 20190238. https://doi.org/10.1098/rspb.2019.0238.

Kuba, Michael J., Tamar Gutnick, and Gordon M. Burghardt. "Learning from Play in Octopus." In *Cephalopod Cognition*, edited by Anne-Sophie Darmaillacq, Ludovic Dickel, and Jennifer Mather, 57–71. Cambridge: Cambridge University Press, 2014. https://doi.org/10.1017/CBO9781139058964.006.

Kuo, Tzu-Hsin, and Chuan-Chin Chiao. "Learned Valuation during Forage Decision-Making in Cuttlefish." *Royal Society Open Science* 7, no. 12 (December 2020): 201602. https://doi.org/10.1098/rsos.201602.

Mattila, Heather R., Gard W. Otis, Lien T. P. Nguyen, Hanh D. Pham, Olivia M. Knight, and Ngoc T. Phan. "Honey Bees (*Apis cerana*) Use Animal Feces as a Tool to Defend Colonies against Group Attack by Giant Hornets (*Vespa soror*)." *PLoS ONE* 15, no. 12 (December 2020): e0242668. https://doi.org/10.1371/journal.pone.0242668.

Mikhalevich, Irina, Russell Powell, and Corina Logan. "Is Behavioural Flexibility Evidence of Cognitive Complexity? How Evolution Can Inform Comparative Cognition." *Interface Focus* 7, no. 3 (June 2017): 20160121. https://doi.org/10.1098/rsfs.2016.0121.

Niven, Jeremy E., and Lars Chittka. "Evolving Understanding of Nervous System Evolution." *Current Biology* 26, no. 20 (October 2016): R937–941. https://doi.org/10.1016/j.cub.2016.09.003.

Pennisi, Elizabeth. "Antisocial Bees Share Genetic Profile with People with Autism." *Science*, July 31, 2017. https://doi.org/10.1126/science.aan7184.

Preston, Elizabeth. "Eight-Armed Underwater Bullies: Watch Octopuses Punch Fish." *New York Times*, December 24, 2020. https://www.nytimes.com/2020/12/24/science/octopus-punch-fish.html.

———. "Meet a Bee with a Very Big Brain." *New York Times*, July 15, 2020. https://www.nytimes.com/2020/09/15/science/bees-brains.html.

Robinson, Gene E. "Darwinian Bee-Keeping: Lessons from the Wild." *Nature* 571, no. 7763 (July 2019): 34–35. https://doi.org/10.1038/d41586-019-02043-3.

Roffet-Salque, Mélanie, Martine Regert, Richard P. Evershed, Alan K. Outram, Lucy J. E. Cramp, Orestes Decavallas, Julie Dunne, et al. "Widespread Exploitation of the Honeybee by Early Neolithic Farmers." *Nature* 527, no. 7577 (November 2015): 226–230. https://doi.org/10.1038/nature15757.

Sampaio, Eduardo, Martim Costa Seco, Rui Rosa, and Simon Gingins. "Octopuses Punch Fishes during Collaborative Interspecific Hunting Events." *Ecology* 102, no. 3 (March 2021): e03266. https://doi.org/10.1002/ecy.3266.

Sayol, Ferran, Miguel Á. Collado, Joan Garcia-Porta, Marc A. Seid, Jason Gibbs, Ainhoa Agorreta, Diego San Mauro, Ivo Raemakers, Daniel Sol, and Ignasi Bartomeus. "Feeding Specialization and Longer Generation Time Are Associated with Relatively Larger Brains in Bees." *Proceedings of the Royal Society B: Biological Sciences* 287, no. 1935 (September 2020): 20200762. https://doi.org/10.1098/rspb.2020.0762.

Shettleworth, Sara J. "Clever Animals and Killjoy Explanations in Comparative Psychology." *Trends in Cognitive Sciences* 14, no. 11 (November 2010): 477–481. https://doi.org/10.1016/j.tics.2010.07.002.

Snell-Rood, Emilie C., and Daniel R. Papaj. "Patterns of Phenotypic Plasticity in Common and Rare Environments: A Study of Host Use and Color Learning in the Cabbage White Butterfly *Pieris rapae*." *American Naturalist* 173, no. 5 (May 2009): 615–631. https://doi.org/10.1086/597609.

Snell-Rood, Emilie C., and Meredith K. Steck. "Behaviour Shapes Environmental Variation and Selection on Learning and Plasticity: Review of Mechanisms and Implications." *Animal Behaviour* 147 (January 2019): 147–156. https://doi.org/10.1016/j.anbehav.2018.08.007.

Thorndike, E. L. "Animal Intelligence: An Experimental Study of the Associative Processes in Animals." *Psychological Review Monograph Supplement* 2 (1898): 1–109.

Weintraub, Karen. "Worker Ants: You Could Have Been Queens." *New York Times*, July 26, 2018. https://www.nytimes.com/2018/07/26/science/ants-genes-queen.html.

Wilke, Carolyn. "The Mirror Test Peers into the Workings of Animal Minds." *The Scientist*, February 21, 2019. https://www.the-scientist.com/news-opinion/the-mirror-test-peers-into-the-workings-of-animal-minds-65497.

Zhou, Aiming, Yuzhe Du, and Jian Chen. "Ants Adjust Their Tool Use Strategy in Response to Foraging Risk." *Functional Ecology* 34, no. 12 (December 2020): 2524–2535. https://doi.org/10.1111/1365-2435.13671.

Zuk, M. *Sex on Six Legs*. (2013) New York, Houghton Mifflin Harcourt.

Zuk, M., J. T. Rotenberry, and R. M. Tinghitella. "Silent Night: Adaptive Disappearance of a Sexual Signal in a Parasitized Population of Field Crickets." *Biology Letters* 2 (2006): 521–524.

9. Talking with the Birds and the Bees. And the Monkeys.

Araya-Salas, M. "Is Birdsong Music? Evaluating Harmonic Intervals in Songs of a Neo-tropical Songbird." *Animal Behaviour* 84 (2012): 309–313.

Balezeau, Fabien, Benjamin Wilson, Guillermo Gallardo, Fred Dick, William Hopkins, Alfred Anwander, Angela D. Friederici, Timothy D. Griffiths, and Christopher I. Petkov. "Primate Auditory Prototype in the Evolution of the Arcuate Fasciculus." *Nature Neuroscience* 23, no. 5 (May 2020): 611–614. https://doi.org/10.1038/s41593 -020-0623-9.

Barham, Lawrence, and Daniel Everett. "Semiotics and the Origin of Language in the Lower Palaeolithic." *Journal of Archaeological Method and Theory* 28, no. 2 (June 2021): 535–579. https://doi.org/10.1007/s10816-020-09480-9.

Bergman, Thore J., Jacinta C. Beehner, Melissa C. Painter, and Morgan L. Gustison. "The Speech-like Properties of Nonhuman Primate Vocalizations." *Animal Behaviour* 151 (2019): 229–237. https://doi.org/10.1016/j.anbehav.2019.02.015.

Bickerton, Derek. "Where Language Comes From." *Harvard University Press Blog* (blog), January 29, 2014. https://harvardpress.typepad.com/hup_publicity/2014/01/more -than-nature-needs-derek-bickerton.html.

Bilger, Hans T., Emily Vertosick, Andrew Vickers, Konrad Kaczmarek, and Richard O. Prum. "Higher-Order Musical Temporal Structure in Bird Song." *Frontiers in Psychology* 12 (March 2021): 629456. https://doi.org/10.3389/fpsyg.2021.629456.

Boë, Louis-Jean, Thomas R. Sawallis, Joël Fagot, Pierre Badin, Guillaume Barbier, Guillaume Captier, Lucie Ménard, Jean-Louis Heim, and Jean-Luc Schwartz. "Which Way to the Dawn of Speech?: Reanalyzing Half a Century of Debates and Data in Light of Speech Science." *Science Advances* 5, no. 12 (December 2019): eaaw3916. https://doi.org/10.1126/sciadv.aaw3916.

Cate, Carel ten, and Michelle Spierings. "Rules, Rhythm and Grouping: Auditory Pattern Perception by Birds." *Animal Behaviour* 151 (May 2019): 249–257. https://doi .org/10.1016/j.anbehav.2018.11.010.

Charles, Krista. "Neanderthal Ears Were Tuned to Hear Speech Just like Modern Humans." *New Scientist*, March 1, 2021. https://www.newscientist.com/article/2269577 -neanderthal-ears-were-tuned-to-hear-speech-just-like-modern-humans/.

Corballis, Michael C. "Crossing the Rubicon: Behaviorism, Language, and Evolutionary Continuity." *Frontiers in Psychology* 11 (April 2020): 653. https://doi.org/10.3389/ fpsyg.2020.00653.

———. "Language Evolution: A Changing Perspective." *Trends in Cognitive Sciences* 21, no. 4 (April 2017): 229–236.

———. "Mental Time Travel, Language, and Evolution." *Neuropsychologia* 134 (November 2019): 107202. https://doi.org/10.1016/j.neuropsychologia.2019 .107202.

Crockford, Catherine, Roman M. Wittig, and Klaus Zuberbühler. "Vocalizing in Chimpanzees Is Influenced by Social-Cognitive Processes." *Science Advances* 3, no. 11 (November 2017): e1701742. https://doi.org/10.1126/sciadv.1701742.

Farias-Virgens, Madza. "Birdsong and Human Language." UC Berkeley Social Science

Matrix, May 23, 2016. https://matrix.berkeley.edu/research-article/birdsong-and -human-language/.

Fishbein, Adam R., Jonathan B. Fritz, William J. Idsardi, and Gerald S. Wilkinson. "What Can Animal Communication Teach Us about Human Language?" *Philosophical Transactions of the Royal Society B: Biological Sciences* 375, no. 1789 (January 2020): 20190042. https://doi.org/10.1098/rstb.2019.0042.

Fitch, W. Tecumseh. "Animal Cognition and the Evolution of Human Language: Why We Cannot Focus Solely on Communication." *Philosophical Transactions of the Royal Society B: Biological Sciences* 375, no. 1789 (January 2020): 20190046. https://doi .org/10.1098/rstb.2019.0046.

Gábor, Anna, Márta Gácsi, Dóra Szabó, Ádám Miklósi, Enikő Kubinyi, and Attila Andics. "Multilevel fMRI Adaptation for Spoken Word Processing in the Awake Dog Brain." *Scientific Reports* 10 (August 2020): 11968. https://doi.org/10.1038/s41598 -020-68821-6.

Ghazanfar, Asif A., Diana A. Liao, and Daniel Y. Takahashi. "Volition and Learning in Primate Vocal Behaviour." *Animal Behaviour* 151 (May 2019): 239–247. https://doi .org/10.1016/j.anbehav.2019.01.021.

Graham, Kirsty E., Catherine Hobaiter, James Ounsley, Takeshi Furuichi, and Richard W. Byrne. "Bonobo and Chimpanzee Gestures Overlap Extensively in Meaning." *PLoS Biology* 16, no. 2 (February 2018): e2004825. https://doi.org/10.1371/journal .pbio.2004825.

Graham, Kirsty E., Claudia Wilke, Nicole J. Lahiff, and Katie E. Slocombe. "Scratching beneath the Surface: Intentionality in Great Ape Signal Production." *Philosophical Transactions of the Royal Society B: Biological Sciences* 375, no. 1789 (January 2020): 20180403. https://doi.org/10.1098/rstb.2018.0403.

Hockett, Charles F. "The Origin of Speech." *Scientific American* 203, no. 3 (September 1960): 88–97.

Hockett, Charles F., and Robert Ascher. "The Human Revolution [and Comments and Reply]." *Current Anthropology* 5, no. 3 (June 1964): 135–168. https://doi.org/10 .1086/200477.

Humphrey, Tasmin, Leanne Proops, Jemma Forman, Rebecca Spooner, and Karen McComb. "The Role of Cat Eye Narrowing Movements in Cat–Human Communication." *Scientific Reports* 10 (October 2020): 16503. https://doi.org/10.1038/s41598 -020-73426-0.

Jackendoff, Ray. "How Did Language Begin?" Linguistic Society of America, August 2006.

Kershenbaum, Arik, Vlad Demartsev, David E. Gammon, Eli Geffen, Morgan L. Gustison, Amiyaal Ilany, and Adriano R. Lameira. "Shannon Entropy as a Robust Estimator of Zipf's Law in Animal Vocal Communication Repertoires." *Methods in Ecology and Evolution* 12, no. 3 (December 2020): 553–564. https://doi.org/10.1111/2041 -210X.13536.

Kwok, Sinead. "The Human–Animal Divide in Communication: Anthropocentric, Posthuman and Integrationist Answers." *Language & Communication* 74 (September 2020): 61–73. https://doi.org/10.1016/j.langcom.2020.06.005.

Newcastle University. "Origins of Human Language Pathway in the Brain at Least 25 Million Years Old." *ScienceDaily*, April 20, 2020. https://www.sciencedaily.com/releases/2020/04/200420125519.htm.

Nieder, Andreas, and Richard Mooney. "The Neurobiology of Innate, Volitional and Learned Vocalizations in Mammals and Birds." *Philosophical Transactions of the Royal Society B: Biological Sciences* 375, no. 1789 (January 2020): 20190054. https://doi.org/10.1098/rstb.2019.0054.

Novack, Miriam A., and Sandra Waxman. "Becoming Human: Human Infants Link Language and Cognition, but What about the Other Great Apes?" *Philosophical Transactions of the Royal Society B: Biological Sciences* 375, no. 1789 (January 2020): 20180408. https://doi.org/10.1098/rstb.2018.0408.

Pereira, André S., Eithne Kavanagh, Catherine Hobaiter, Katie E. Slocombe, and Adriano R. Lameira. "Chimpanzee Lip-Smacks Confirm Primate Continuity for Speech-Rhythm Evolution." *Biology Letters* 16, no. 5 (May 2020): 20200232. https://doi.org/10.1098/rsbl.2020.0232.

Piattelli, Michela. "'Language Is Our Rubicon': Friedrich Max Müller's Quarrel with Hensleigh Wedgwood." *Publications of the English Goethe Society* 85, no. 2–3 (2016): 98–109. https://doi.org/10.1080/09593683.2016.1224511.

Pinker, Steven, and Ray Jackendoff. "The Faculty of Language: What's Special about It?" *Cognition* 95, no. 2 (March 2005): 201–236. https://doi.org/10.1016/j.cognition.2004.08.004.

Scharff, Constance, Mirjam Knörnschild, and Erich D. Jarvis. "Vocal Learning and Spoken Language: Insights from Animal Models with an Emphasis on Genetic Contributions." In *Human Language: From Genes and Brains to Behavior*, edited by Peter Hagoort, 657–685. (2019) Cambridge, MA: MIT Press.

Searcy, William A. "Animal Communication, Cognition, and the Evolution of Language." *Animal Behaviour* 151 (May 2019): 203–205. https://doi.org/10.1016/j.anbehav.2019.03.001.

Searcy, William A., and Stephen Nowicki. "Birdsong Learning, Avian Cognition and the Evolution of Language." *Animal Behaviour* 151 (May 2019): 217–227. https://doi.org/10.1016/j.anbehav.2019.01.015.

Speck, Bretta, Sara Seidita, Samuel Belo, Samuel Johnson, Caley Conley, Camille Desjonquères, and Rafael L. Rodríguez. "Combinatorial Signal Processing in an Insect." *American Naturalist* 196, no. 4 (October 2020): 406–413. https://doi.org/10.1086/710527.

Starr, Michelle. "Study Confirms 'Slow Blinks' Really Do Work to Communicate with Your Cat." *ScienceAlert*, October 8, 2020. https://www.sciencealert.com/you-can-build-a-rapport-with-your-cat-by-blinking-real-slow.

Suzuki, Toshitaka N., Michael Griesser, and David Wheatcroft. "Syntactic Rules in Avian Vocal Sequences as a Window into the Evolution of Compositionality." *Animal Behaviour* 151 (May 2019): 267–274. https://doi.org/10.1016/j.anbehav.2019.01.009.

Suzuki, Toshitaka N., and Klaus Zuberbühler. "Animal Syntax." *Current Biology* 29, no. 14 (July 2019): R669–671. https://doi.org/10.1016/j.cub.2019.05.045.

Szathmáry, Eörs, and John Maynard Smith. "The Major Evolutionary Transitions." *Nature* 374 (March 16, 1995): 227–232. https://doi.org/10.1038/374227a0.

Underwood, Emily. "Birdsong Not Music, after All." *Science*, August 15, 2012. https://www.science.org/news/2012/08/birdsong-not-music-after-all.

Vernes, Sonja C., and Gerald S. Wilkinson. "Behaviour, Biology and Evolution of Vocal Learning in Bats." *Philosophical Transactions of the Royal Society B: Biological Sciences* 375, no. 1789 (January 2020): 20190061. https://doi.org/10.1098/rstb.2019.0061.

Wade, Nicholas. "Early Voices: The Leap to Language." *New York Times*, July 15, 2003. https://www.nytimes.com/2003/07/15/science/early-voices-the-leap-to-language.html.

Wilson, A. N. "Why I Believe Again." *New Statesman*, April 2, 2009. https://www.newstatesman.com/religion/2009/04/conversion-experience-atheism.

Yong, Ed. "Can Humans Understand Chimps?" *The Atlantic*, August 15, 2017. https://www.theatlantic.com/science/archive/2017/08/can-humans-understand-chimps/536826/.

Zuberbühler, Klaus. "Evolutionary Roads to Syntax." *Animal Behaviour* 151 (May 2019): 259–265. https://doi.org/10.1016/j.anbehav.2019.03.006.

10. The Faithful Coucal

Ah-King, Malin, and Ingrid Ahnesjö. "The 'Sex Role' Concept: An Overview and Evaluation." *Evolutionary Biology* 40, no. 4 (December 2013): 461–470. https://doi.org/10.1007/s11692-013-9226-7.

Ahnesjö, Ingrid, Jaelle C. Brealey, Katerina P. Günter, Ivain Martinossi-Allibert, Jennifer Morinay, Mattias Siljestam, Josefine Stångberg, and Paula Vasconcelos. "Considering Gender-Biased Assumptions in Evolutionary Biology." *Evolutionary Biology* 47, no. 1 (March 2020): 1–5. https://doi.org/10.1007/s11692-020-09492-z.

Barnett, Rosalind C., and Caryl Rivers. "We've Studied Gender and STEM for 25 Years. The Science Doesn't Support the Google Memo." *Vox*, August 11, 2017. https://www.vox.com/2017/8/11/16127992/google-engineer-memo-research-science-women-biology-tech-james-damore.

Borgerhoff Mulder, Monique, and Cody T. Ross. "Unpacking Mating Success and Testing Bateman's Principles in a Human Population." *Proceedings of the Royal Society B: Biological Sciences* 286, no. 1908 (August 2019): 20191516. https://doi.org/10.1098/rspb.2019.1516.

Broughton, Darcy E., Robert E. Brannigan, and Kenan R. Omurtag. "Sex and Gender: You Should Know the Difference." *Fertility and Sterility* 107, no. 6 (June 2017): 1294–1295. https://doi.org/10.1016/j.fertnstert.2017.04.012.

Cameron, Deborah. "Evolution, Language and the Battle of the Sexes: A Feminist Linguist Encounters Evolutionary Psychology." *Australian Feminist Studies* 30, no. 86 (October 2015): 351–358. https://doi.org/10.1080/08164649.2016.1148097.

———. "Sex/Gender, Language and the New Biologism." *Applied Linguistics* 31, no. 2 (May 2010): 173–192. https://doi.org/10.1093/applin/amp022.

Damore, James. "Google's Ideological Echo Chamber: How Bias Clouds Our Thinking about Diversity and Inclusion," July 2017.

Davis, Jac T. M., and Melissa Hines. "How Large Are Gender Differences in Toy Preferences? A Systematic Review and Meta-analysis of Toy Preference Research." *Archives of Sexual Behavior* 49, no. 2 (February 2020): 373–394. https://doi.org/10.1007/s10508-019-01624-7.

Duchesne, Annie, Belinda Pletzer, Marina A. Pavlova, Meng-Chuan Lai, and Gillian Einstein. "Editorial: Bridging Gaps between Sex and Gender in Neurosciences." *Frontiers in Neuroscience* 14 (June 2020): 561. https://doi.org/10.3389/fnins.2020.00561.

Eliot, Lise, and Sarah S. Richardson. "Sex in Context: Limitations of Animal Studies for Addressing Human Sex/Gender Neurobehavioral Health Disparities." *Journal of Neuroscience* 36, no. 47 (November 2016): 11823–11830. https://doi.org/10.1523/JNEUROSCI.1391-16.2016.

Griffiths, Paul. "Sex Is Real." Aeon. September 21, 2020. https://aeon.co/essays/the-existence-of-biological-sex-is-no-constraint-on-human-diversity.

Haig, David. "The Inexorable Rise of Gender and the Decline of Sex: Social Change in Academic Titles, 1945–2001." *Archives of Sexual Behavior* 33, no. 2 (April 2004): 87–96. https://doi.org/10.1023/B:ASEB.0000014323.56281.0d.

Harden, Kathryn Paige. "The Science of Terrible Men." Aeon. March 11, 2021. https://aeon.co/essays/what-do-we-do-with-the-science-of-terrible-men.

Janicke, Tim, Ines K. Häderer, Marc J. Lajeunesse, and Nils Anthes. "Darwinian Sex Roles Confirmed across the Animal Kingdom." *Science Advances* 2, no. 2 (February 2016): e1500983. https://doi.org/10.1126/sciadv.1500983.

Janssen, Diederik F. "Know Thy Gender: Etymological Primer." *Archives of Sexual Behavior* 47, no. 8 (November 2018): 2149–2154. https://doi.org/10.1007/s10508-018-1300-x.

Lee, Cynthia. "I'm a Woman in Computer Science. Let Me Ladysplain the Google Memo to You." *Vox*, August 11, 2017. https://www.vox.com/the-big-idea/2017/8/11/16130452/google-memo-women-tech-biology-sexism.

Lipshutz, Sara E., and Kimberly A. Rosvall. "Neuroendocrinology of Sex-Role Reversal." *Integrative and Comparative Biology* 60, no. 3 (September 2020): 692–702. https://doi.org/10.1093/icb/icaa046.

Martin, Robert D. "Intersex: Life in the Overlap Zone." *Psychology Today*, September 24, 2019.

———. "No Substitute for Sex." *Psychology Today*, August 20, 2019. https://www.psychologytoday.com/us/blog/how-we-do-it/201908/no-substitute-sex.

Molteni, Megan, and Adam Rogers. "The Actual Science of James Damore's Google Memo." *Wired*, August 15, 2017. https://www.wired.com/story/the-pernicious-science-of-james-damores-google-memo/.

Ogle, Derek H., and Kevin F. Schanning. "Usage of 'Sex' and 'Gender.'" *Fisheries* 37, no. 6 (June 2012): 271–272. https://doi.org/10.1080/03632415.2012.687265.

Orr, Teri J., and Virginia Hayssen. "The Female Snark Is Still a Boojum: Looking toward the Future of Studying Female Reproductive Biology." *Integrative and Com-*

parative Biology 60, no. 3 (September 2020): 782–795. https://doi.org/10.1093/icb/icaa091.

Pogrebna, Ganna, Andrew J. Oswald, and David Haig. "Female Babies and Risk-Aversion: Causal Evidence from Hospital Wards." *Journal of Health Economics* 58 (March 2018): 10–17. https://doi.org/10.1016/j.jhealeco.2017.12.006.

Rippon, Gina. "The Trouble with Girls?" *The Psychologist* 29, no. 12 (December 2016): 918–922.

Ristvedt, Stephen L. "The Evolution of Gender." *JAMA Psychiatry* 71, no. 1 (January 2014): 13–14. https://doi.org/10.1001/jamapsychiatry.2013.3199.

Rivkis, Nora. Comment on "If Gender Roles Are Cultural Constructs, How Can There Be Gender Roles in Animal Species?" Quora, October 28, 2016. https://www.quora.com/If-gender-roles-are-cultural-constructs-how-can-there-be-gender-roles-in-animal-species.

Sadedin, Suzanne. Comment on "What Do Feminists Think of Distinct Gender Roles in Other Species, for Example, in Chickens?" Quora. Accessed August 10, 2017. https://www.quora.com/What-do-feminists-think-of-distinct-gender-roles-in-other-species-for-example-in-chickens.

———. Comment on "What Do Scientists Think about the Biological Claims Made in the Anti-diversity Document Written by a Google Employee in August 2017?" Quora. Accessed August 10, 2017. https://www.quora.com/What-do-scientists-think-about-the-biological-claims-made-in-the-anti-diversity-document-written-by-a-Google-employee-in-August-2017/answer/Suzanne-Sadedin.

Safari, Ignas, and Wolfgang Goymann. "The Evolution of Reversed Sex Roles and Classical Polyandry: Insights from Coucals and Other Animals." *Ethology* 127, no. 1 (January 2021): 1–13. https://doi.org/10.1111/eth.13095.

Şahin, Özlem, and Nur Soylu Yalcinkaya. "The Gendered Brain: Implications of Exposure to Neuroscience Research for Gender Essentialist Beliefs." *Sex Roles* 84, no. 2 (May 2021): 522–535. https://doi.org/10.1007/s11199-020-01181-7.

Schwartz, Jay. "Is Gender Unique to Humans?" *Sapiens*, November 29, 2018. https://www.sapiens.org/culture/gender-identity-nonhuman-animals/.

Shahvisi, Arianne. "Nesting Behaviours during Pregnancy: Biological Instinct, or Another Way of Gendering Housework?" *Women's Studies International Forum* 78 (January 2020): 102329. https://doi.org/10.1016/j.wsif.2019.102329.

Shansky, Rebecca M. "Sex Differences in Behavioral Strategies: Avoiding Interpretational Pitfalls." *Current Opinion in Neurobiology* 49 (April 2018): 95–98. https://doi.org/10.1016/j.conb.2018.01.007.

Stevens, Sean, and Jonathan Haidt. "The Google Memo: What Does the Research Say about Gender Differences?" *Heterodox* (blog), August 10, 2017. https://heterodoxacademy.org/blog/the-google-memo-what-does-the-research-say-about-gender-differences/.

Tang-Martínez, Zuleyma. "Rethinking Bateman's Principles: Challenging Persistent Myths of Sexually Reluctant Females and Promiscuous Males." *The Journal of Sex Research* 53, no. 4–5 (2016): 532–559. https://doi.org/10.1080/00224499.2016.1150938.

Xu, Min, Xiuling Liang, Jian Ou, Hong Li, Yue-jia Luo, and Li Hai Tan. "Sex Differences in Functional Brain Networks for Language." *Cerebral Cortex* 30, no. 3 (March 2020): 1528–1537. https://doi.org/10.1093/cercor/bhz184.

11. Protect and Defend

Abbott, Jessica. "Self-Medication in Insects: Current Evidence and Future Perspectives." *Ecological Entomology* 39, no. 3 (June 2014): 273–280. https://doi.org/10.1111/een.12110.

Barelli, Claudia, and Michael A. Huffman. "Leaf Swallowing and Parasite Expulsion in Khao Yai White-Handed Gibbons (*Hylobates lar*), the First Report in an Asian Ape Species." *American Journal of Primatology* 79, no. 3 (March 2017): e22610. https://doi.org/10.1002/ajp.22610.

Barry, Colleen. "Highly Contagious Coronavirus Variants Powering Another Surge in Europe." *Los Angeles Times*, March 6, 2021.

Billing, J., and P. W. Sherman. "Antimicrobial Functions of Spices: Why Some Like It Hot." *Quarterly Review of Biology* (1998) 73: 3–49.

Boillat, Madlaina, Pierre-Mehdi Hammoudi, Sunil Kumar Dogga, Stéphane Pagès, Maged Goubran, Ivan Rodriguez, and Dominique Soldati-Favre. "Neuroinflammation-Associated Aspecific Manipulation of Mouse Predator Fear by *Toxoplasma gondii*." *Cell Reports* 30, no. 2 (January 2020): 320–334.e6. https://doi.org/10.1016/j.celrep.2019.12.019.

Bos, Nick, Liselotte Sundström, Siiri Fuchs, and Dalial Freitak. "Ants Medicate to Fight Disease." *Evolution* 69, no. 11 (November 2015): 2979–2984. https://doi.org/10.1111/evo.12752.

Bromham, Lindell, Alexander Skeels, Hilde Schneemann, Russell Dinnage, and Xia Hua. "There Is Little Evidence That Spicy Food in Hot Countries Is an Adaptation to Reducing Infection Risk." *Nature Human Behaviour* 5, no. 7 (July 2021): 878–891. https://doi.org/10.1038/s41562-020-01039-8.

Bromhan, Lindell. "Why Do Hot Countries Have Spicy Food?" *Behind the Paper* (blog), SocialSciences.Nature, February 4, 2021. https://socialsciences.nature.com/posts/why-do-hot-countries-have-spicy-food?badge_id=569-nature-human-behaviour.

Chapuisat, Michel, Anne Oppliger, Pasqualina Magliano, and Philippe Christe. "Wood Ants Use Resin to Protect Themselves against Pathogens." *Proceedings of the Royal Society B: Biological Sciences* 274, no. 1621 (August 2007): 2013–2017. https://doi.org/10.1098/rspb.2007.0531.

Choisy, Marc, and Jacobus C. de Roode. "The Ecology and Evolution of Animal Medication: Genetically Fixed Response versus Phenotypic Plasticity." *The American Naturalist* 184, no. S1 (August 2014): S31–46. https://doi.org/10.1086/676928.

Costes-Thiré, Morgane, Juan J. Villalba, Hervé Hoste, and Cécile Ginane. "Increased Intake and Preference for Tannin-Rich Sainfoin (*Onobrychis viciifolia*) Pellets by Both Parasitized and Non-parasitized Lambs after a Period of Conditioning." *Applied Animal Behaviour Science* 203 (March 2018): 11–18. https://doi.org/10.1016/j.applanim.2018.02.015.

Coyne, Jerry. "Sea Slug Regrows Entire Body from Just the Decapitated Head, or 'Autotomy with Kleptoplasty.'" *Why Evolution Is True* (blog), March 9, 2021. https://whyevolutionistrue.com/2021/03/09/sea-slug-regrows-entire-body-from-just-the-decapitated-head-or-autotomy-with-kleptoplasty/.

Doherty, Jean-François. "When Fiction Becomes Fact: Exaggerating Host Manipulation by Parasites." *Proceedings of the Royal Society B: Biological Sciences* 287, no. 1936 (October 2020): 20201081. https://doi.org/10.1098/rspb.2020.1081.

Frank, Erik T., Marten Wehrhahn, and K. Eduard Linsenmair. "Wound Treatment and Selective Help in a Termite-Hunting Ant." *Proceedings of the Royal Society B: Biological Sciences* 285, no. 1872 (February 2018): 20172457. https://doi.org/10.1098/rspb.2017.2457.

Fredericksen, Maridel A., Yizhe Zhang, Missy L. Hazen, Raquel G. Loreto, Colleen A. Mangold, Danny Z. Chen, and David P. Hughes. "Three-Dimensional Visualization and a Deep-Learning Model Reveal Complex Fungal Parasite Networks in Behaviorally Manipulated Ants." *Proceedings of the National Academy of Sciences* 114, no. 47 (November 21, 2017): 12590–12595. https://doi.org/10.1073/pnas.1711673114.

Greene, Alexander M., Prateep Panyadee, Angkhana Inta, and Michael A. Huffman. "Asian Elephant Self-Medication as a Source of Ethnoveterinary Knowledge among Karen Mahouts in Northern Thailand." *Journal of Ethnopharmacology* 259 (2020): 112823. https://doi.org/10.1016/j.jep.2020.112823.

Grens, Kerry. "How Mice Forget to Be Afraid." *The Scientist*, May 1, 2020. https://www.the-scientist.com/the-literature/how-mice-forget-to-be-afraid-67442.

Hardy, Karen. "Paleomedicine and the Evolutionary Context of Medicinal Plant Use." *Revista Brasileira de Farmacognosia* 31 (February 2021): 1–15. https://doi.org/10.1007/s43450-020-00107-4.

———. "Paleomedicine and the Use of Plant Secondary Compounds in the Paleolithic and Early Neolithic." *Evolutionary Anthropology* 28, no. 2 (April 2019): 60–71. https://doi.org/10.1002/evan.21763.

Hardy, Karen, Stephen Buckley, and Michael Huffman. "Doctors, Chefs or Hominin Animals? Non-edible Plants and Neanderthals." *Antiquity* 90, no. 353 (October 2016): 1373–1379. https://doi.org/10.15184/aqy.2016.134.

Herbison, Ryan E. H. "Lessons in Mind Control: Trends in Research on the Molecular Mechanisms behind Parasite-Host Behavioral Manipulation." *Frontiers in Ecology and Evolution* 5 (September 2017): 102. https://doi.org/10.3389/fevo.2017.00102.

Huffman, Michael A. "Animal Self-Medication and Ethno-medicine: Exploration and Exploitation of the Medicinal Properties of Plants." *Proceedings of the Nutrition Society* 62, no. 2 (May 2003): 371–381. https://doi.org/10.1079/PNS2003257.

———. "Self-Medicative Behavior in the African Great Apes: An Evolutionary Perspective into the Origins of Human Traditional Medicine." *BioScience* 51, no. 8 (August 2001): 651–661. https://doi.org/10.1641/0006-3568(2001)051[0651:SMBITA]2.0.CO;2.

Lefèvre, Thierry, Lindsay Oliver, Mark D. Hunter, and Jacobus C. de Roode. "Evidence for Trans-generational Medication in Nature." *Ecology Letters* 13, no. 12 (December 2010): 1485–1493. https://doi.org/10.1111/j.1461-0248.2010.01537.x.

Lisonbee, Larry D., Juan J. Villalba, Fred D. Provenza, and Jeffery O. Hall. "Tannins and Self-Medication: Implications for Sustainable Parasite Control in Herbivores." *Behavioural Processes* 82, no. 2 (October 2009): 184–189. https://doi.org/10.1016/j .beproc.2009.06.009.

Lopes, Patricia C. "We Are Not Alone in Trying to Be Alone." *Frontiers in Ecology and Evolution* 8 (June 2020): 172. https://doi.org/10.3389/fevo.2020.00172.

Makin, Douglas F., Burt P. Kotler, Joel S. Brown, Mario Garrido, and Jorge F. S. Menezes. "The Enemy Within: How Does a Bacterium Inhibit the Foraging Aptitude and Risk Management Behavior of Allenby's Gerbils?" *American Naturalist* 196, no. 6 (December 2020): 717–729. https://doi.org/10.1086/711397.

Manson, Jessamyn S., Michael C. Otterstatter, and James D. Thomson. "Consumption of a Nectar Alkaloid Reduces Pathogen Load in Bumble Bees." *Oecologia* 162, no. 1 (January 2010): 81–89. https://doi.org/10.1007/s00442-009-1431-9.

McGrew, William M. "In Search of the Last Common Ancestor: New Findings on Wild Chimpanzees." *Philosophical Transactions of the Royal Society B* (2010) 365: 3267–3276. doi:10.1098/rstb.2010.0067.

Mitoh, Sayaka, and Yoichi Yusa. "Extreme Autotomy and Whole-Body Regeneration in Photosynthetic Sea Slugs." *Current Biology* 31, no. 5 (March 2021): R233–234. https:// doi.org/10.1016/j.cub.2021.01.014.

Morrogh-Bernard, H. C., I. Foitová, Z. Yeen, P. Wilkin, R. de Martin, L. Rárová, K. Doležal, W. Nurcahyo, and M. Olšanský. "Self-Medication by Orang-Utans (*Pongo pygmaeus*) Using Bioactive Properties of *Dracaena cantleyi*." *Scientific Reports* 7 (2017): 16653. https://doi.org/10.1038/s41598-017-16621-w.

Moutinho, Sofia. "Why Cats Are Crazy for Catnip." *Science*, January 20, 2021. https:// www.science.org/news/2021/01/why-cats-are-crazy-catnip.

Poissonnier, Laure-Anne, Mathieu Lihoreau, Tamara Gomez-Moracho, Audrey Dussutour, and Jerome Buhl. "A Theoretical Exploration of Dietary Collective Medication in Social Insects." *Journal of Insect Physiology* 106 (April 2018): 78–87. https://doi .org/10.1016/j.jinsphys.2017.08.005.

Poyet, M., P. Eslin, O. Chabrerie, S. M. Prud'homme, E. Desouhant, and P. Gibert. "The Invasive Pest *Drosophila suzukii* Uses Trans-generational Medication to Resist Parasitoid Attack." *Scientific Reports* 7 (2017): 43696. https://doi.org/10 .1038/srep43696.

Reiber, Chris, and Janice Moore. "Synergies That Work: Evolution, Epidemiology, and New Insights." *Annals of Epidemiology* 20, no. 10 (October 2010): 725–728. https:// doi.org/10.1016/j.annepidem.2010.06.006.

Reiber, Chris, Eric C. Shattuck, Sean Fiore, Pauline Alperin, Vanessa Davis, and Janice Moore. "Change in Human Social Behavior in Response to a Common Vaccine." *Annals of Epidemiology* 20, no. 10 (October 2010): 729–733. https://doi.org/10.1016/j .annepidem.2010.06.014.

Roode, Jacobus C. de, and Mark D. Hunter. "Self-Medication in Insects: When Altered Behaviors of Infected Insects Are a Defense Instead of a Parasite Manipulation." *Current Opinion in Insect Science* 33 (June 2019): 1–6. https://doi.org/10.1016/j.cois .2018.12.001.

Roode, Jacobus C. de, Thierry Lefèvre, and Mark D. Hunter. "Self-Medication in Ani-

mals." *Science* 340, no. 6129 (April 2013): 150–151. https://doi.org/10.1126/science
.1235824.

Servick, Kelly. "Brain Parasite May Strip Away Rodents' Fear of Predators—Not Just of
Cats." *Science*, January 14, 2020. https://doi.org/10.1126/science.aba8985.

Sherman, Paul W., and J. Billing. "Darwinian Gastronomy: Why We Use Spices: Spices
Taste Good because They Are Good for Us." *BioScience* (1999) 49: 453–463.

Simone-Finstrom, Michael D., and Marla Spivak. "Increased Resin Collection after
Parasite Challenge: A Case of Self-Medication in Honey Bees?" *PLoS ONE* 7, no. 3
(March 2012): e34601. https://doi.org/10.1371/journal.pone.0034601.

Singer, Michael S., Kevi C. Mace, and Elizabeth A. Bernays. "Self-Medication as Adaptive Plasticity: Increased Ingestion of Plant Toxins by Parasitized Caterpillars."
PLoS ONE 4, no. 3 (March 10, 2009): e4796. https://doi.org/10.1371/journal.pone
.0004796.

Singer, Michael S., Peri A. Mason, and Angela M. Smilanich. "Ecological Immunology
Mediated by Diet in Herbivorous Insects." *Integrative and Comparative Biology* 54,
no. 5 (June 2014): 913–921. https://doi.org/10.1093/icb/icu089.

Spivak, Marla, and Martha Gilliam. "Facultative Expression of Hygienic Behaviour of
Honey Bees in Relation to Disease Resistance." *Journal of Apicultural Research* 32, no.
3–4 (1993): 147–157. https://doi.org/10.1080/00218839.1993.11101300.

Spivak, Marla, Michael Goblirsch, and Michael Simone-Finstrom. "Social-Medication in Bees: The Line between Individual and Social Regulation." *Current Opinion in Insect Science* 33 (June 2019): 49–55. https://doi.org/10.1016/j.cois.2019.02
.009.

Spivak, Marla, Rebecca Masterman, Rocco Ross, and Karen A. Mesce. "Hygienic Behavior in the Honey Bee (*Apis mellifera* L.) and the Modulatory Role of Octopamine."
Journal of Neurobiology 55, no. 3 (June 2003): 341–354. https://doi.org/10.1002/neu
.10219.

Suárez-Rodríguez, Monserrat, and Constantino Macías Garcia. "An Experimental
Demonstration That House Finches Add Cigarette Butts in Response to Ectoparasites." *Journal of Avian Biology* 48, no. 10 (October 2017): 1316–1321. https://doi
.org/10.1111/jav.01324.

Tao, Leiling, Kevin M. Hoang, Mark D. Hunter, and Jacobus C. de Roode. "Fitness Costs
of Animal Medication: Antiparasitic Plant Chemicals Reduce Fitness of Monarch
Butterfly Hosts." *Journal of Animal Ecology* 85, no. 5 (September 2016): 1246–1254.
https://doi.org/10.1111/1365-2656.12558.

Troisi, Alfonso. "Fear of COVID-19: Insights from Evolutionary Behavioral Science." *Clinical Neuropsychiatry* 17, no. 2 (2020): 72–75. https://doi.org/10.36131/
CN20200207.

Uenoyama, Reiko, Tamako Miyazaki, Jane L. Hurst, Robert J. Beynon, Masaatsu
Adachi, Takanobu Murooka, Ibuki Onoda, et al. "The Characteristic Response of
Domestic Cats to Plant Iridoids Allows Them to Gain Chemical Defense against
Mosquitoes." *Science Advances* 7, no. 4 (January 2021): eabd9135. https://doi.org/10
.1126/sciadv.abd9135.

Ventura-Cordero, J., P. G. González-Pech, P. R. Jaimez-Rodriguez, G. I. Ortiz-Ocampo,
C. A. Sandoval-Castro, and J. F. J. Torres-Acosta. "Feed Resource Selection of Criollo

Goats Artificially Infected with *Haemonchus contortus*: Nutritional Wisdom and Prophylactic Self-Medication." *Animal* 12, no. 6 (2018): 1269–1276. https://doi.org/10.1017/S1751731117002634.

Villalba, J. J., and F. D. Provenza. "Self-Medication and Homeostatic Behaviour in Herbivores: Learning about the Benefits of Nature's Pharmacy." *Animal* 1, no. 9 (2007): 1360–1370. https://doi.org/10.1017/S1751731107000134.

Villalba, Juan J., James Miller, Eugene D. Ungar, Serge Y. Landau, and John Glendinning. "Ruminant Self-Medication against Gastrointestinal Nematodes: Evidence, Mechanism, and Origins." *Parasite* 21 (2014): 31. https://doi.org/10.1051/parasite/2014032.

Webster, J. P. "The Effect of *Toxoplasma gondii* on Animal Behavior: Playing Cat and Mouse." *Schizophrenia Bulletin* 33 (2007): 752–756. doi: 10.1093/schbul/sbl073.

Wu, Katherine J. "Your Cat Isn't Just Getting High off Catnip." *New York Times*, January 20, 2021. https://www.nytimes.com/2021/01/20/science/catnip-mosquito-repellent.html.

Yong, Ed. "How the Zombie Fungus Takes over Ants' Bodies to Control Their Minds." *The Atlantic*, November 14, 2017. https://www.theatlantic.com/science/archive/2017/11/how-the-zombie-fungus-takes-over-ants-bodies-to-control-their-minds/545864/.

Index

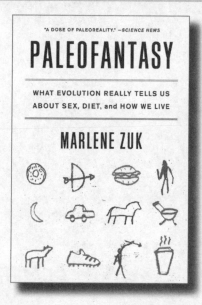
"With . . . evidence from recent genetic and anthropological research, [Marlene Zuk] offers a dose of paleoreality."

—Erin Wayman, *Science News*

"Much-needed. . . . Ms. Zuk's nutritionally rich scientific fodder will certainly bring intellectual benefits far greater than those provided by the pseudoscientific confections with which we are so often tempted."

—Cordelia Fine, *Wall Street Journal*

"[Zuk] ably presents a sceptical and light-hearted view of a long list of palaeofantasies and supposed solutions."

—John Hawks, *Nature*

W. W. NORTON & COMPANY
Celebrating a Century of Independent Publishing